SOLUTIONS MANUAL
PAUL L. GAUS

TO ACCOMPANY

BASIC INORGANIC CHEMISTRY / THIRD EDITION

F. ALBERT COTTON

W. T. Doherty-Welch Foundation
Distinguished Professor of Chemistry
Texas A and M University
College Station, Texas, USA

GEOFFREY WILKINSON

Emeritus Professor of Inorganic Chemistry
Imperial College of Science, Technology, and Medicine
London SW7 2AY
England

PAUL L. GAUS

Professor of Chemistry
The College of Wooster
Wooster, Ohio, USA

JOHN WILEY & SONS, INC.
NEW YORK • CHICHESTER • BRISBANE • TORONTO • SINGAPORE

ISBN 0-471-51808-5

Printed in the United States of America

10 9 8 7 6 5 4

With love and gratitude to my wife Madonna

1-1

The terms endothermic and exothermic are defined by reference to equation 1-2.1. For endothermic processes, ΔH is positive; for exothermic processes, ΔH is negative.

1-3

(a) $H_2O(s) \rightarrow H_2O(g)$

(b) $C_6H_6(l) \rightarrow C_6H_6(g)$

(c) $Cl(g) + e^- \rightarrow Cl^-(g)$

(d) $Na(g) \rightarrow Na^+(g) + e^-$

1-5

(a)

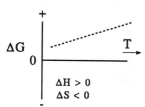
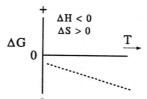
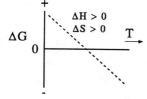
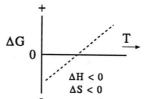

(b)

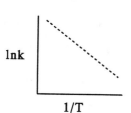

1-7

Entropy is defined to be zero for a perfect crystalline solid at 0 K.

1-9

Typical graphs are presented below, specifically for the case where $k = 1 \times 10^{-3}$ M^{-1} s^{-1}.

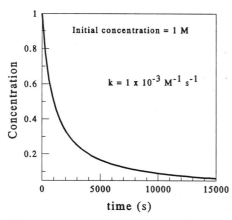
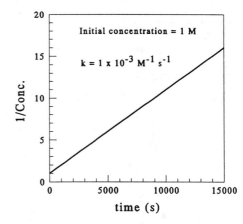

1-11

The N-N bond in F_2N-NF_2 is obviously weaker than the N-N bond in H_2N-NH_2. Cleavage of these N-N bonds produces a radical, the stability of which depends on the substituents at N. Fluorine substitution in H_2N-NH_2 to give F_2N-NF_2 weakens the N-N bond, because the F_2N radical is more stable:

$$F_2N\text{-}NF_2 \rightarrow 2F_2N \qquad \Delta H = 80 \text{ kJ mol}^{-1}$$
$$H_2N\text{-}NH_2 \rightarrow 2H_2N \qquad \Delta H = 160 \text{ kJ mol}^{-1}$$

Compare these values with the N-N single bond energy given in Table 1-1.

1-13

(a) $1/2 H_2 + 1/2 N_2 + Cl_2 \rightarrow HNCl_2$

$\Delta H_f°[HNCl_2] = -[(E_{N-H} + 2 \times E_{N-Cl})$
$\qquad\qquad\qquad - (1/2 \times E_{H-H} + 1/2 \times E_{N \equiv N} + E_{Cl-Cl})]$
$\qquad = -[391 = 2 \times 193) - (1/2 \times 436 + 1/2 \times 946 + 242)]$
$\qquad = +156 \text{ kJ mol}^{-1}$

(b) The method of using bond energies to estimate ΔH can be used only for reactions in which all reactants and products are in the gaseous state.

2

(c) $Cl_2 + N_2 + H_2 \rightarrow Cl_2N-NH_2$

$$\Delta H_f°[Cl_2NNH_2] = -[(2 \times E_{N-Cl} + 2 \times E_{N-H} + E_{N-N})$$
$$- (E_{H-H} + E_{Cl-Cl} + E_{N \equiv N})]$$
$$= -[(2 \times 193 + 2 \times 391 + 160)$$
$$- (436 + 242 + 946)]$$
$$= +296 \text{ kJ mol}^{-1}$$

1-15

$CO + H_2O \rightarrow CO_2 + H_2$

$$\Delta H = -[(2 \times E_{C=O} + E_{H-H}) - (2 \times E_{O-H} + E_{C \equiv O})]$$
$$= -[(2 \times 695 + 436) - (2 \times 467 + 1073)]$$
$$= +181 \text{ kJ mol}^{-1}$$

1-17

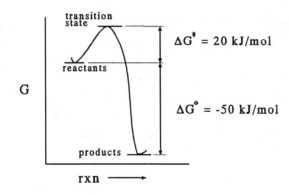

1-19

(a) $E° = 3.04 \text{ V} + 1.36 \text{ V} = 4.40 \text{ V}$

(b) $E° = -0.53 \text{ V} + 1.61 \text{ V} = 1.08 \text{ V}$

CHAPTER 2

A - Review

2-1A
Equation 2-1.1, developed by Rydberg, represents the sets of emission lines in the spectrum of the hydrogen atom.

2-3A
This is Planck's equation, $E = h\nu$. The energy of a quantum of electromagnetic radiation is directly proportional to the frequency of the electromagnetic radiation. The constant of proportionality is h, Planck's constant.

2-5A
de Broglie proposed that any particle in motion is characterized by a wavelength, as given in equation 2-2.1. The wave nature of the electron was first demonstrated by the experiments of Davisson and Germer, namely that a beam of electrons can undergo diffraction.

2-7A
This information is given in Table 2-2. The quantum number n can take integer values from 1 to infinity, although values higher than 7 are not chemically meaningful. The azimuthal quantum number l takes values in integer steps from 0 up to (n-1). The quantum number m_l ranges in integer steps from $-l$ to $+l$.

2-9A
See Figure 2-8.

2-11A
An orbital may have appreciable amplitude in regions of space close to the nucleus, and to the extent that it does, the electron is stabilized. The orbitals which are closer to the nucleus (as shown in Figure 2-7) are, therefore, more stable. In particular, the s-type orbitals of any given principal shell penetrate the space of p and d orbitals of the same shell. This causes the s orbital to shield the orbitals of other types from the nuclear charge. The relative degree of penetration of the orbitals of any given principal

4

shell (s > p > d > f) consequently determines their relative energies
(s < p < d < f).

2-13A
Hund's first rule states that the ground state electron configuration
is the one with the maximum spin multiplicity. See section 2-4 for
the application of this rule when listing ground state electron
configurations for the elements.

2-15A
See equation 2-8.1 and Table 2-5.

2-17A
Mulliken's electronegativities are calculated using the average of
the electron attachment enthalpy and the ionization enthalpy:
$\chi = 1/2(\Delta H_{EA} + \Delta H_{ion})$.

2-19A
The ionic radius that is assigned to an element may depend in general
on such factors as (a) the oxidation state of the element, (b) the
coordination number of the element in the compound at hand, (c) the
nature and identity of any counter ions, (d) the presence of unfilled
d-orbitals (Chapter 23), and (e) the degree of covalency in a
particular substance.

B - Additional Exercises

2-1B
Substituting Z = 2 into equation 2-1.4, we realize (from equation 2-
1.5) that for He^+ we have:

$$\bar{v} = 4R\left[\frac{1}{m^2} - \frac{1}{n^2}\right]$$

where R = 109,678 cm^{-1}. The following values are obtained.

Line	$\bar{v}$ (cm^{-1})

For the "Lyman" series (m = 1),
    the first line (from n = 2):   329,036
    the last line, or series limit (from n = ∞):   438,712
For the "Balmer" series (m = 2),
    the first line (from n = 3):   60,932
    the last line, or series limit (from n = ∞):   109,680
For the "Paschen" series (m = 3),
    the first line (from n = 4):   21,328
    the last line, or series limit (from n = ∞):   48,744

2-3B

(a)
$$\lambda = \frac{6.6256 \times 10^{-27} \text{ erg s}}{(9.1091 \times 10^{-28} g)(10^6 \text{ cm s}^{-1})}$$

$$\lambda = 7.27 \times 10^{-6} \text{ cm} = 727 \text{ Å}$$

(b) Let m = 2.00 × 10^2 g.

$$\lambda = \frac{6.6256 \times 10^{-27} \text{ erg s}}{(200 \text{ g})(10^3 \text{ cm s}^{-1})}$$

$$\lambda = 3.31 \times 10^{-32} \text{ cm} = 3.31 \times 10^{-24} \text{ Å}$$

2-5B

The two lines correspond to the transitions from states where n = ∞ to m = 4 ($\bar{v}$ = 6855 cm^{-1}) and from states where n = ∞ to m = 5 (n = 4387 cm^{-1}). See problem 2-1B, and use equation 2-1.5.

2-7B

This requires use of de Broglie's matter-wave equation 2-2.1.
Proceeding as in problem 2-3B, and letting $\lambda = 6.0 \times 10^{-8}$ cm, we
have:

$$v = \frac{6.6256 \times 10^{-27} \text{ erg s}}{(9.1 \times 10^{-28}\text{g})(6.0 \times 10^{-8} \text{ cm})}$$

$$v = 1.2 \times 10^{8} \text{ cm s}^{-1}$$

2-9B

From equation 2-7.5 we get:

$$z^* = \frac{(\chi - 0.744)r^2}{0.359}$$

Z	χ	r($\mathring{A}$)	z^* (relative)
19	0.91	1.98	1.8
20	1.04	1.78	2.6
31	1.82	1.30	5.1
32	2.02	1.22	5.3

It is sensible that trends in physical properties such as
electronegativity and covalent radius should parallel changes in
effective nuclear charge. As the effective nuclear charge increases,
size decreases and the tendency to attract electrons increases.

2-11B

Each diagram should have the shape indicated by Figure 2-6. The
cross section of each orbital should be shaded with an intensity that
varies according to the function $r^2R(r)^2$, as shown in Figure 2-7B.
At distances r where nodes are indicated, the shading should be
lightest (technically zero), and the shading should
become darker in proportion to the value of the function $r^2R(r)^2$.

2-13B

Using the information in Table 2-5:

    (a) $MnSO_4 \cdot 4H_2O$, 5 unpaired electrons

    (b) $CuSO_4 \cdot 5H_2O$, 1 unpaired electron

    (c) $(NH_4)_2Fe(SO_4)_2 \cdot 6H_2O$, 4 or 5 unpaired electrons

    (d) $[Cr(NH_3)_6](NO_3)_3$, 3 unpaired electrons

    (e) $[Cu(NH_3)_4]SO_4 \cdot 3H_2O$, 1 unpaired electrons

    (f) $[Co(NH_3)_6]Cl_3$, 0 unpaired electrons

CHAPTER 3

A - Review

3-1A
Covalent bonding is accomplished *via* the overlap of orbitals. By the overlap criterion of bond strength, bond strength is high when the signs of the wave functions match and when the area of overlap is great. Under these circumstances, the cross terms in the function $(\phi_A + \phi_B)^2$ add constructively to the electron density between the nuclei, as shown by curve 3 in Figure 3-16.

3-3A
See Figure 3-19. The bond order is 1 for H_2; it is 0 for He_2.

3-5A
Orbitals which are designated σ, π and δ have, respectively, zero, one and two nodes in planes that are perpendicular to the internuclear axis.

3-7A
Bond order is defined in section 3-5 to be the difference between the number of bonding pairs of electrons (n_b) and the number of antibonding pairs of electrons (n_a). It is also equal to 1/2 the difference between the number of electrons in bonding MO's and the number of electrons in antibonding MO's.

3-9A
The energy level diagrams are quite similar, NO having one less electron to be placed in the MO diagram (Figure 3-26). The bond order for NO is 2.5. NO^+ has one electron less than NO, and its bond order is three; it is isoelectronic with CO (Figure 3-27), and like CO, its energy level diagram reflects the polarity in the molecule.

3-11A

Ground States	Valence States

Be: $1s^2 2s^2$ $1s^2 2s^1 2p^1$

B: $1s^2 2s^2 2p^1$ $1s^2 2s^1 (2p_x)^1 (2p_y)^1$

C: $1s^2 2s^2 2p^2$ $1s^2 2s^1 (2p_x)^1 (2p_y)^1 (2p_z)^1$

N: $1s^2 2s^2 2p^3$ $1s^2 2s^2 (2p_x)^1 (2p_y)^1 (2p_z)^1$

3-13A

See Figures 3-2 through 3-8.

3-15A

The most obvious reason for this is geometric. The atomic p orbitals are disposed at 90° to one another, and are thus unable to accommodate geometries such as the tetrahedron (109.5°).

3-17A

B_2H_6 is an electron deficient molecule. The structure is shown in Figure 3-33, and there are two types of bonds in the molecule: two three-center, two-electron bonds of the BHB bridges, and four two-center, two-electron bonds of the terminal BH groups.

3-19A

In molecular bromine, each atom is isoelectronic with elemental Kr. The interatomic distance in krypton should therefore be similar to the nonbonded, intermolecular distance between bromine of separate molecules. The intramolecular, or covalent radius of bromine must obviously be smaller than either one of the above van der Waals radii.

3-21A

Because of the high effective nuclear charge on a fluorine atom, there is a substantial separation in the energies of the atomic s and p orbitals. This precludes s-p mixing in F_2. The effective nuclear charge on a lithium atom, however, is smaller, and s-p mixing is possible in Li_2.

B - Additional Exercises

3-1B

Refer to Figure 3-26.

Molecule or Ion	MO Configuration	Bond Order
O_2^+	$[\sigma_1]^2[\sigma_2]^2[\sigma_3]^2[\pi_1]^4[\pi_2]^1$	2.5
O_2	$[\sigma_1]^2[\sigma_2]^2[\sigma_3]^2[\pi_1]^4[\pi_2]^2$	2.0
O_2^-	$[\sigma_1]^2[\sigma_2]^2[\sigma_3]^2[\pi_1]^4[\pi_2]^3$	1.5
O_2^{2-}	$[\sigma_1]^2[\sigma_2]^2[\sigma_3]^2[\pi_1]^4[\pi_2]^4$	1.0

The number of unpaired electrons is: O_2^+ (1), O_2 (2), O_2^- (1), and O_2^{2-} (0). As the bond order increases, the bond length should decrease.

3-3B

In $(CH_3)_2S$, the C-S-C angle is 99°.
In $(CH_3)_2SO$, the C-S-C angle is 97°, and the C-S-O angle is 107°.

3-5B

This molecule contains Al-Br-Al bridge groups, as well as terminal Al-Br groups. The structure is analogous to that shown in structures 3-XIV or 3-XV. Each Al-Br-Al bridge may be described with the three-center bond approach discussed for diborane, except that the bridge bond system is not electron deficient. Thus, four electrons are to be placed into an energy level diagram similar to that in Figure 3-35 (two electrons into a bonding three-center MO, and two electrons into the nonbonding three-center MO), giving a three-center, four-electron bond system. The four terminal Al-Br groups are well described as two-center, two-electron bonds.

3-7B

We expect gaseous GeF_2 to be bent, since the Ge atom is of the type AB_2E. The crystalline material has an extended structure in which bent (angle F-Ge-F = 85°) F_2Ge units are linked by bridging of the following type:

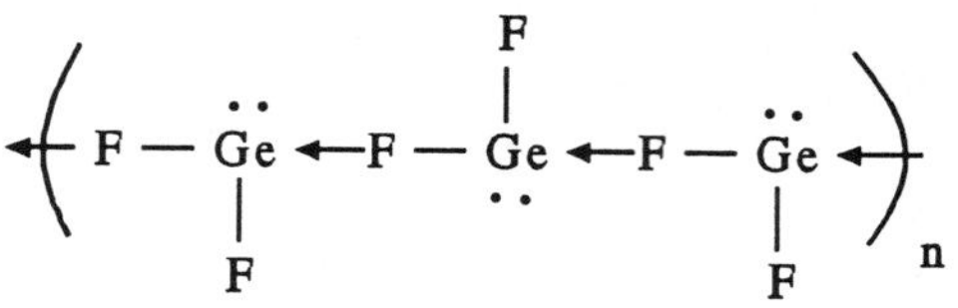

In this fashion, the valence of each Ge atom becomes four in the solid, since the Ge atom in each GeF_2 unit falls into the AB_3E classification. The bridging in the crystalline substance completes the valence requirements of Ge.

3-9B

The Lewis diagrams are:

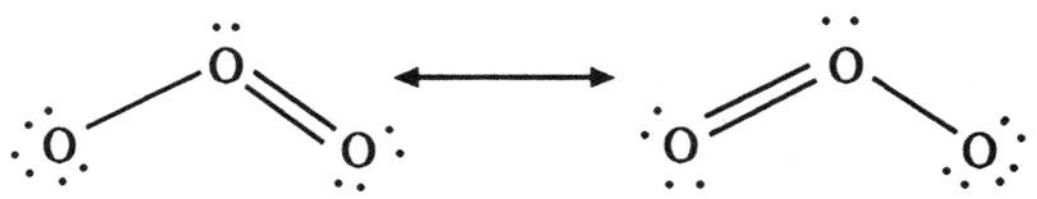

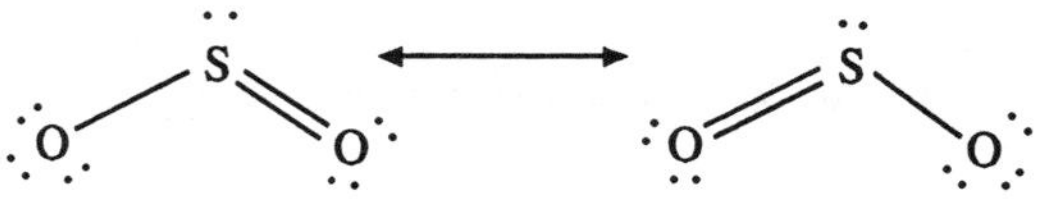

The larger central S atom of SO_2 allows the lone pair sufficient room so as to alleviate the sort of repulsion that collapses the bond angle in O_3. The S atom in SO_2 and the central O atom in O_3 are both sp^2 hybridized and fall into the AB_2E classification.

3-11B

Each of these should have the trigonal planar geometry typical of AB_3 systems.

3-13B

This is an octahedral system, and the occupancy equals six. The hybridization is d^2sp^3, although more will be said on this in Chapter 6.

3-15B

SO_3 is isostructural and isoelectronic with NO_3^-, and the π-bond system in each is delocalized over three linkages.

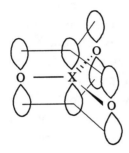

NO^+ has two π bonds.

3-17B

The sum of the single bond covalent radii are:

Si-O, 1.83 Å; P-O, 1.76 Å; S-O, 1.70 Å; Cl-O, 1.65 Å

These ions thus have bonds that are shorter (by 0.20, 0.22, 0.21, and 0.19 Å, respectively) than the simple sum of the single bond covalent radii, suggesting some significant amount of $d\pi$-$p\pi$ double bond character in each case.

3-19B

(a) Molecules and ions from Question 3-18B that are electronically saturated: a, c, d, e, f, g, h, i, j, k, p, q, r, and s.

(b) Molecules and ions from Question 3-18B that are coordinatively saturated: a, e, g, j, and q.

C - Questions From the Literature of Inorganic Chemistry

3-1C.

This is the anion $[TeF_5]^-$, and the oxidation state of Te is formally Te^{IV}. The Lewis diagram contains bonds to the five fluorine atoms and a lone pair of electrons on the central Te atom. This is an AB_5E system, and the occupancy is 6. The Te can be considered to be d^2sp^3 hybridized. The lone pair is placed into one of the six positions of an octahedron, and the geometry of the anion is derived from a square pyramid. The basal fluorine atoms are bent out of the plane of the Te atom (and towards the apical F atom) because of repulsion from the lone pair on Te. There also are interactions between adjacent $[TeF_5]^-$ units, and this perturbs the geometry further, as discussed in the article.

3-3C.

Although the $AlMe_3$ groups appear to be crowded, the normal geometry of the tetrahedron is found for the S atom. We can then consider this to be a typical AB_4 system for S, an AB_4 system for each Al atom, and an AB_2E_2 system for each O atom.

CHAPTER 4

A - Review

4-1A

Both repulsive (equation 4-2.5) and attractive (equation 4-2.3) forces must be considered, as in equation 4-2.6. Additionally, the lattice geometry is to be considered.

4-3A

The Born expression for the non-Coulomb repulsive energy is equation 4-2.5. The constant B had been eliminated in reaching equation 4-2.9. The repulsive forces represented by equation 4-2.5 do not follow simple coulombic (varying according to $1/r^2$) behavior (as do the attractive forces), rather they vary according to $1/r^n$.

4-5A

Pauling proposed that, for ions which are isoelectronic, the ratio of their sizes (ionic radii) should be inversely proportional to the ratio of their effective nuclear charges. The larger the effective nuclear charge, the smaller the ion.

4-7A

Close packing of spheres in a layer (or plane) is accomplished when the spheres touch to give interlocking equilateral triangles, each sphere being touched symmetrically by six other spheres. The space between spheres (unoccupied space) is minimized.

4-9A

The difference between ccp and hcp arrays lies in the way a third layer of close packed spheres is added to the two layers of Structure 4-1. Hexagonal closest packing is accomplished by ABABABAB... stacking. Cubic closest packing is obtained by placing a third close packed layer over top of the two in Structure 4-I so as to align the spheres of the third layer over those holes of the first layer that were not covered by spheres of the second layer. This is then an ABCABCABC... stacking of layers. See also Figures 4-5 and 4-6.

4-11A

This is the mineral spinel. Two different cations occupy tetrahedral and octahedral holes in a cubic close packed array of oxide ions.

4-13A

In the fluorite structure, the cations are arranged in ccp fashion, and the anions occupy tetrahedral holes. In anitfluroite structures, the anions are packed in a ccp array, and the cations occupy the tetrahedral holes.

4-15A

Corundum and ilmenite both have an hcp array of oxide anions with aluminum (in corundum) or iron and titanium (in ilmenite) ions in only two-thirds of the octahedral holes.

In perovskite, the oxide ions plus the large cations jointly form a ccp array, and the small titanium ions occupy octahedral holes formed by the oxide ions alone.

B - Additional Exercises

4-1B

We can begin the analysis with any cation or anion in the line. Each term in the infinite series has for its numerator the number two, because of the symmetry of the array: each ion to the left of the chosen one has its counterpart (of like charge) an equal distance to the right. Thus the first term in the series is 2/d, where d is the interionic distance. The second term is -2/2d, accounting for two ions each at distance 2d from the chosen one, and each having the same sign as the chosen one. This last interaction is repulsive, and the sign of the term is negative. Two ions at distance 3d account for the third term: 2/3d. The series is written as follows:

$$2/d - 2/2d + 2/3d - 2/4d + 2/5d - 2/6d + 2/7d + \ldots = M/d$$

$$M = 1.34$$

4-3B

As shown in Figure 4-1, the edge of a unit cell traverses twice the sum of the radii of the two ions. Thus 4.21 Å = $2(r_+ + r_-)$, and the sum of the radii is 2.10 Å. The ratio of the two radii is the inverse ratio of their effective nuclear charges:

$$\frac{r_+}{r_-} = \frac{8-4.15}{12-4.15} = 0.490$$

Thus we have two equations in two unknowns, and we can solve for the separate values of the radii: $r_- = 1.41$ Å and $r_+ = 0.69$ Å.

4-5B

Equation 4-3.1 may be rewritten to allow calculations of lattice energies in kJ mol^{-1}, using r_o values in Å:

$$U = -1389 \times \frac{M_{NaCl}}{r_o}\left(1-\frac{1}{n}\right)Z^2$$

Hence U = -3.95×10^3 kJ mol^{-1} for MgO.

We continue:

Mg(s) → Mg(g), $\Delta H_1 = \Delta H_{vap} = 150.2$ kJ

$1/2 O_2(g)$ → O(g), $\Delta H_2 = 1/2 E_{O=O} = 249$ kJ

Mg(g) → Mg^{2+}(g) + 2e$^-$, $\Delta H_3 = \Delta H_{ion}(1) + \Delta H_{ion}(2)$
                             $= 2.19 \times 10^3$ kJ

O(g) + 2e$^-$ → O^{2-}(g), $\Delta H_4 = ?$

Mg^{2+}(g) + O^{2-}(g) → MgO(s), $\Delta H_5 = U$

Mg(s) + $1/2 O_2$(g) → MgO(s), $\Delta H_6 = \Delta H_f[MgO(s)] = -602$ kJ

Solving for ΔH_4:   759 kJ mol^{-1}

4-7B

Crystals that have a higher compressibility are more able to undergo a shortening of interionic distances. The lattice energy repulsive terms for such substances must therefore be lower than in crystals which have low compressibilities.

4-9B

Na_2S has the antifluorite structure with anions in the ccp array and all tetrahedral holes filled with cations. Since the cation to anion ratio in this formula is 2:1, there must be two tetrahedral holes for every anion.

NaCl has a structure with anions arranged in a ccp fashion, all octahedral holes being filled with cations. Therefore, there must be one octahedral hole for every anion, since the ratio of cation to anion in this substance is 1:1.

In $CdCl_2$, the anions form a ccp array, with cations filling only half of the octahedral holes. There must be one octahedral hole for every anion, since the ratio of cation to anion is 1:2.

4-11B

See Figure 4-8.

4-13B

For LiF, the ratio $r_-/r_+ = 1.35/0.60 = 2.25$, and from Table 4-3 we predict the NaCl structure (coordination number 6).

For CsI, the ratio $r_-/r_+ = 2.16/1.69 = 1.28$, and from Table 4-3 we predict the CsCl structure (coordination number 8).

In CsCl, the anion is small enough relative to the cation that a large coordination number is possible.

4-15B

Zinc blende has the same structure as diamond.

4-17B

The geometry of the unit cell provides us with the following relationship:

$$r_+ + r_- = \frac{\sqrt{3}}{2} \times a$$

which gives us the value: $a = 4.30 \ \overset{\circ}{A}$.

4-19B

For NaCl:

$$
\begin{array}{rl}
\text{8 vertex Cl:} & 8 \times 1/8 = 1 \text{ Cl} \\
\text{12 edge Na:} & 12 \times 1/4 = 3 \text{ Na} \\
\text{6 face Cl:} & 6 \times 1/2 = 3 \text{ Cl} \\
\text{1 internal Na:} & 1 \times 1 = 1 \text{ Na}
\end{array}
$$

Adding gives us the formula Na_4Cl_4, or NaCl.

For ZnS:

$$
\begin{array}{rl}
\text{8 vertex S:} & 8 \times 1/8 = 1 \text{ S} \\
\text{6 face S:} & 6 \times 1/2 = 3 \text{ S} \\
\text{4 internal Zn:} & 4 \times 1 = 4 \text{ Zn}
\end{array}
$$

Adding gives us the formula Zn_4S_4, or ZnS.

For $CaTiO_3$:

$$
\begin{array}{rl}
\text{8 vertex Ca:} & 8 \times 1/8 = 1 \text{ Ca} \\
\text{6 face O:} & 6 \times 1/2 = 3 \text{ O} \\
\text{1 internal Ti:} & 1 \times 1 = 1 \text{ Ti}
\end{array}
$$

Adding gives us the formula $CaTiO_3$.

For TiO_2:

$$8 \text{ vertex Ti: } 8 \times 1/8 = 1 \text{ Ti}$$
$$4 \text{ face O: } 4 \times 1/2 = 2 \text{ O}$$
$$2 \text{ internal O: } 2 \times 1 = 2 \text{ O}$$
$$1 \text{ internal Ti: } 1 \times 1 = 1 \text{ Ti}$$

Adding gives the formula Ti_2O_4, or TiO_2.

For CsCl:

$$8 \text{ vertex Cl: } 8 \times 1/8 = 1 \text{ Cl}$$
$$1 \text{ internal Cs: } 1 \times 1 = 1 \text{ Cs}$$

Adding gives us the formula CsCl.

For CaF_2:

$$8 \text{ vertex Ca: } 8 \times 1/8 = 1 \text{ Ca}$$
$$6 \text{ face Ca: } 6 \times 1/2 = 3 \text{ Ca}$$
$$8 \text{ internal F: } 8 \times 1 = 8 \text{ F}$$

Adding gives us the formula Ca_4F_8, or CaF_2.

A - Review

5-1A

Oxide is hydrolyzed readily, according to equation 5-2.1.

5-3A

Most of the main group elements form discrete oxoanions of one kind or another, as do many of the transition elements. See section 5-3 in particular. Polymeric or extended oxoanions are also formed by main group elements such as B, Si and P, as well as many transition metals, as discussed in section 5-4.

5-5A

(a) $Cr_2O_7^{2-}$ is shown in structure 5-XXIV.

(b) $Si_2O_7^{6-}$ has a structure similar to that of dichromate, although the Si-O-Si angle varies from one substance to another, and the discrete anion is not common.

(c) $B_2O_5^{4-}$ is a planar ion with two AB_3-type boron atoms, one oxygen atom bridging the two boron atoms.

5-7A

Zeolites are framework minerals derived from silica by substitution of some Si by Al^{3+} ions. The substitution of a formally Si^{4+} ion by Al^{3+} leaves each unit negatively charged, and charge neutrality is preserved in the mineral by the presence of various exchangeable cations such as Na^+ or K^+. The extent of substitution varies with the preparation of the material, as does the cavity size and the extent of hydration. See section 5-4 for further details.

5-9A

See structure 5-XXXI and 5-XXXII.

CHAPTER 5

B - Additional Exercises

5-1B

MgO is a base anhydride that is insoluble in water, but which reacts with acid, as in equation 5-2.3.

B_2O_3 is obtained by fusion of boric acid ($B(OH)_3$). Boron more naturally forms polynuclear oxoanions such as that found in borax, and illustrated in structure 5-XXVII. More information is available in Chapter 12, especially Figure 12-1.

Sb_2O_3 is the oxide of antimony in its highest oxidation state. It is generally obtained by the action of nitric acid on the metal, giving hydrated solids that are soluble only upon conversion to the sulfide (Sb_2S_3), which is sufficiently basic so as to dissolve in HCl. Sb_2O_3 should be considered an acidic oxide due to its reaction as in equation 5-2.5.

The oxide of Si is quartz, whose structure is discussed in section 5-4, along with other silicates. SiO_2 is considered to be a purely acidic oxide, as discussed in chapter 15.

5-3B

See Figure 24-1. Note the presence of three types of ethoxide ligands: terminal, doubly-bridging, and triply-bridging. In the hydrolysis reaction, one or more ethoxide ligands is replaced by an OH^- ligand, with the release of ethanol:

$$[Ti(OEt)_4]_4 + H_2O \rightarrow Ti_4(OEt)_{15}(OH) + EtOH$$

Further hydrolysis eventually gives TiO_2.

5-5B

(a) See Figure 12-1.

(b) This is the empirical unit of a polymer, shown in Figure 12-1 as $[(BO_2)^-]_n$.

(c) This is the anion $[B_5O_6(OH)_4]^-$, shown in Figure 12-1.

5-7B

The cyanide ion, CN^-, is isoelectronic with N_2 and with CO. An MO diagram similar to those shown in Figure 3-27 must therefore be drawn. Of course the energies of the atomic orbitals and the molecular orbitals will be somewhat different for CN^- than for either CO or N_2, because of the different effective nuclear charges that operate in each of these molecules or ions. In each case, though, the highest occupied MO is σ_3, and the lowest unoccupied MO is the doubly degenerate set π^2. The π^2 set is empty and can accept electron density from the metal d orbitals through the sort of overlap shown below:

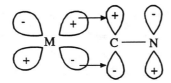

5-9B

$$\left[:\ddot{O}\!-\!\overset{\displaystyle ..}{X}\!-\!\ddot{O}: \atop \underset{:O:}{|} \right]^{-} \qquad \left[\overset{\displaystyle :\ddot{O}:}{\underset{:O:}{\overset{|}{:\ddot{O}\!-\!X\!-\!\ddot{O}:}}} \right]^{-}$$

5-11B

This reaction yields the hydrous oxide, $Cr_2O_3 \cdot nH_2O$.

C - Questions from the Literature of Inorganic Chemistry

5-1C

(a) Two methods are reported for the synthesis of the combined oxides. Fusion (1700 °C) for the titanite:

$$2Sc_2O_3 + 3TiO_2 + Ti \rightarrow 4ScTiO3$$

and for the vanadite, either fusion (1500 °C in a platinum capsule):

$$Sc_2O_3 + V_2O_3 \rightarrow 2ScVO3$$

or fusion to a higher oxide:

$$Sc_2O_3 + V_2O_5 \rightarrow 2ScVO4$$

followed by reduction with hydrogen:

$$ScVO_4 + H_2 \rightarrow ScVO_3 + H_2O$$

(b) The oxide Sc_2O_3 should be considered to contain Sc^{III}, whose outermost electron configuration is the same as that of Ar. The oxidation state of scandium in both the titanite and the vanadite is also Sc^{III}. The magnetic susceptibility of $ScVO_3$ (μ_{eff} = 2.98 BM) suggests two unpaired electrons (see Table 2-5 of the text) on V^{3+}. This is then an ion with outermost electron configuration $[Ar]3d^2$. Titanium in $ScTiO_3$ should be considered most correctly to be the $[Ar]3d^1$ ion (Ti^{III}), although the stoichiometry of the preparation seems to influence the magnetic susceptibility of the product.

(c) See the answer to B above.

(d) The X-ray intensity data for the titanite, the vanadite and the oxide Sc_2O_3 are all reported to be the same, suggesting that these compounds adopt the same face centered cubic structure. See references in the article for the details of the structures.

CHAPTER 6

A - Review

6-1A

Coordination Number	Geometry	Example
Two	linear	$[Cu(CN)_2]^-$
Three	planar	$[HgI_3]^-$
	pyramidal	$[SnCl_3]^-$
Four	tetrahedral	$Ni(CO)_4$
	square	$[PtCl_4]^{2-}$
Five	trigonal bipyramid	$[CuCl_5]^{3-}$
	square pyramid	$[Ni(CN)_5]^{3-}$
Six	octahedron	$W(CO)_6$
	trigonal prism	
Seven	pentagonal bipyramid	
	face-capped trigonal prisms	
Eight	See Figure 6-1	
Nine	See Figure 6-2	

6-3A

(a) A tetragonal distortion of an octahedron involves either a shortening or a lengthening of the bond distances along one L-M-L axis, as shown in structure 6-IV. The extreme case of lengthening in a tetragonal fashion gives a square, four-coordinate complex.

(b) Structure 6-VII is derived from an octahedron by a rhombic distortion.

(c) Trigonal distortion of an octahedron results in structure 6-VIII.

6-5A

   acetylacentonate: See the section on bidentate ligands.

   ethylenediamine: See the section on bidentate ligands.

diethylenetriamine: See the section on tridentate ligands.

       $EDTA^{4-}$: See Figure 6-4.

6-7A

Type of Isomer		Example
1. ionization isomers	1a.	$[Coen_2(NO_2)Cl]SCN$
	1b.	$[Coen_2(SCN)Cl]NO_2$
	1c.	$[Coen_2(SCN)(NO_2)]Cl$
2. linkage isomers	2a.	$[Co(NH_3)_5-NCS]^{2+}$
	2b.	$[Co(NH_3)_5-SCN]^{2+}$
3. coordination isomers	3a.	$[Co(NH_3)_6][Cr(CN)_6]$
	3b.	$[Co(CN)_6][Cr(NH_3)_6]$

6-9A

The N stepwise formation constants K_1, K_2, ... K_N are defined as in equation 6-4.1. Each successive equation in 6-4.1 represents the addition of one more ligand to the preceeding complex in the series. The overall formation constants β_1, β_2, ...β_N are defined as in equation 6-4.2, and each such constant represents a product of the appropriate stepwise constants, as in equations 6-4.3 (i.e. for β_3) or 6-4.4 (representing the general case).

6-11A

The chelate effect is demonstrated by comparing the values of the overall formation constants for two amine complexes of Co^{III}: one involving monodentate ligands ($[Co(NH_3)_6]^{3+}$; $\beta_6 = 10^{34}$) and one involving bidentate ligands ($[Co(en)_3]^{3+}$; $\beta_3 = 10^{49}$). The complex with the bidentate ligand en is more stable (by a factor of 10^{15}!)

than the complex with monodentate ligands. This is an entropy
effect. The probability of coordination of the whole en ligand is
high, once half of it is bound to the metal.

6-13A

The kinetic terms *inert* and *labile* refer to the rate at which a
complex undergoes substitution. Labile complexes react quickly;
inert ones react slowly. Of course the terms "quickly" and "slowly"
are relative ones.

The thermodynamic terms *stable* and *unstable* are applied to a complex
depending on the free energy change associated with a reaction of the
complex, and they are not terms which indicate rate. An unstable
complex is one which spontaneously undergoes reaction (as defined by
thermodynamic spontanaeity, i.e. $\Delta G < 0$), and the definition does
not involve the rate at which the reaction proceeds.

6-15A

(a) Where the solvent can also serve as an entering ligand, the rate
law for substitution is reduced to a simple first order one,
regardless of the mechanism, because of the high and essentially
constant concentration of the solvent. See reactions 6-5.5 and
6-5.6.

(b) Ion pair formation obscures the molecularity of the rate
determining step in a substitution mechanism, giving second order
kinetics regardless of the associative or dissociative nature of the
rate determining step of the mechanism.

(c) The conjugate-base mechanism must be considered (as well as
simple substitution involving OH^-) when the experimental rate law
contains $[OH^-]$.

6-17A

The alternate explanation is that the conjugate-base mechanism
operates, and that the acceleration of the reaction is due to the

extraordinary ability of the NH_2^- ligand to promote cleavage of the bond to the ligand trans to itself.

6-19A

The trans effect of a ligand is its ability to promote rapid substitution of the ligand in the position trans to itself.

6-21A

Taube has demonstrated that the inner sphere mechanism operates in a number of reactions. The proof of mechanism is sometimes difficult, with the exception of one particular situation. The inner sphere mechanism must operate in reactions where all of the following things are true: (1) a bridging ligand is transferred from one metal center to the other, (2) the bridging ligand among the reactants is bound to an inert metal center, and the other metal reactant is substitution labile, and (3) the bridging ligand among the products is bound to a new metal center which is substitution inert because of the change in oxidation state it has sustained.

B - Additional Exercises

6-1B

The following process places the basal group B_1 into the apical position formerly occupied by B_5:

6-3B

The octahedral geometry can yield only two isomers with this formula: *cis* and *trans* isomers.

A trigonal prismatic geometry allows for the existence of three isomers:

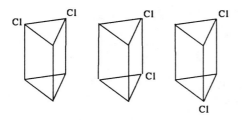

6-5B

Two geometric isomers exist, *cis* and *trans*.
Two optical isomers exist because the *cis* isomer is chiral.

6-7B

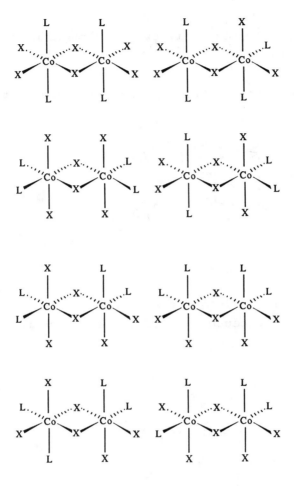

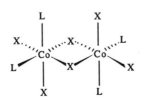

6-9B

The substitution reactions of square platinum complexes are associative. Factors which make for stable Pt-X bonds within the reactant PtX_4^{2-} (i.e. factors which lead to high thermodynamic stability) are also those that lead to stability in the transition state (i.e. factors which make for low activation barriers).

6-11B

(a) $Fe(CN)_6]^{4-} + [Fe(CN)_6]^{3-} \rightarrow [Fe(CN)_6]^{3-} + [Fe(CN)_6]^{4-}$

$[Coen_3]^{3+} + [Coen_3]^{2+} \rightarrow [Coen_3]^{2+} + [Coen_3]^{3+}$

(b) $[Fe(H_2O)_6]^{3+} + [Fe(H_2O)_6]^{2+} \rightarrow [Fe(H_2O)_6]^{2+} + [Fe(H_2O)_6]^{3+}$

$[Cr(phen)_3]^{2+} + [Cr(phen)_3]^{3+} \rightarrow [Cr(phen)_3]^{3+} + [Cr(phen)_3]^{2+}$

(c) $[Rh(phen)_3]^{3+} + [Rh(phen)_3]^{2+} \rightarrow [Rh(phen)_3]^{2+} + [Rh(phen)_3]^{3+}$

$[Ru(phen)_3]^{2+} + [Ru(phen)_3]^{3+} \rightarrow [Ru(phen)_3]^{3+} + [Ru(phen)_3]^{2+}$

6-13B

(a) dichlorodiammineplatinum(II)
(b) chloropentaamminerhodium(III) chloride
(c) hexaamminecobalt(III) nitrate
(d) tetraaquocobalt(II) sulfate
(e) tetraamminediaquocobalt(III) tetrafluoroborate
(f) hexaaquoiron(II) bromide
(g) sodium hexacyanoferrate(III) dihydrate
(h) sodium hexacyanoferrate(II)
(i) tetracarbonylnickel(0)
(j) tetraamminecopper(II) sulfate
(k) bis(ethylenediamine)platinum(II) perchlorate
(l) acetatobromochlorodiamminecobalt(III)
(m) hexaaquochromium(II) chloride
(n) tris(ethylenediamine)cobalt(III) sulfate
(o) sodium hydridotrimethoxoborate(III)

(p) tetrakispyridineplatinum(II) tetrachloroplatinate(II)

(q) sodium hexachloropalladate(IV)

(r) tetraethylammonium hexacyanochromate(III)

(s) trisphenanthrolinenickel(II) perchlorate

(t) nitropentaamminecobalt(III) sulfate

(u) chloronitrobis(ethylenediamine)cobalt(III) thiocyanate

(v) tetramethylammonium chloropentacarbonyltungstate(0)

(w) hexaaquochromium(III) chloride

(x) acetylacetonatochloroammineplatinum(II)

6-15B

(a)

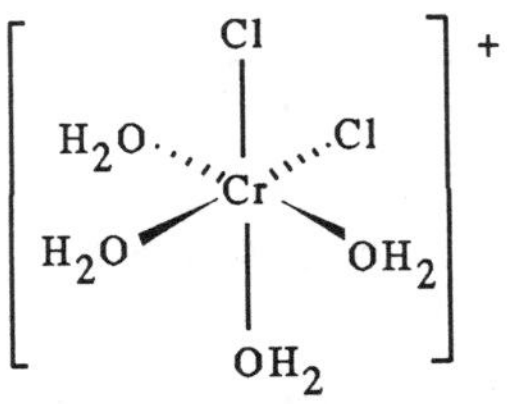

cis 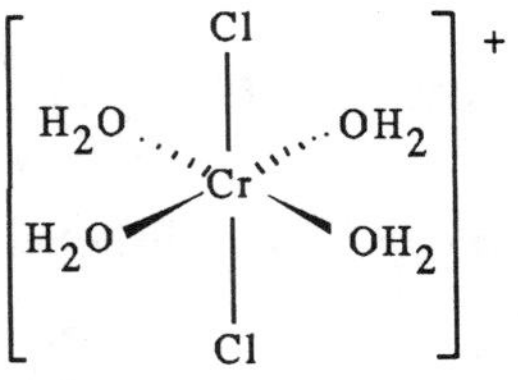   trans

(b)

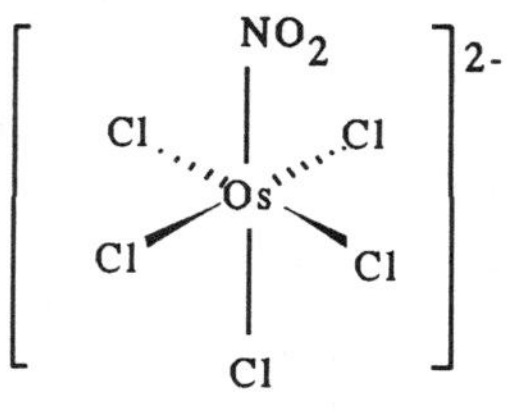

nitro 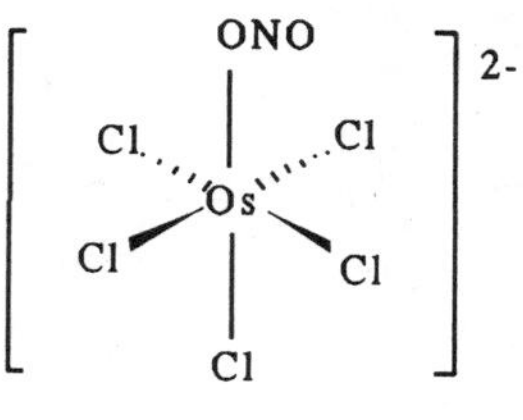   nitrito

(c)

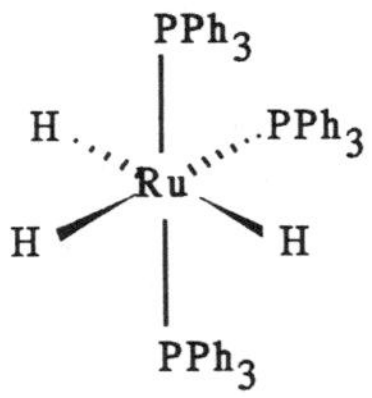

meridional 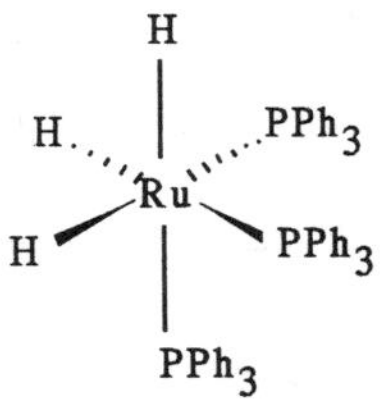   facial

(d)

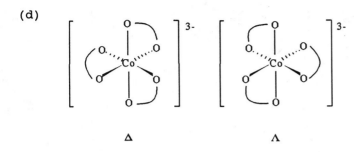

(e)

$$\left[\begin{array}{c} NH_3 \\ H_3N\cdots \overset{}{\underset{}{Co}}\cdots NH_3 \\ H_3N \quad\quad Cl \\ NH_3 \end{array}\right]^{2+}$$

(f)

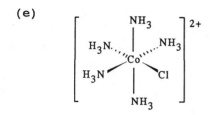

(g)

$$\left[\begin{array}{c} Br\cdots Pt\cdots Br \\ Br \quad\quad Br \end{array}\right]^{2-}$$

(h)

$$\left[\begin{array}{c} NH_3 \\ H_3N\cdots \overset{}{\underset{}{Ru}}\cdots NH_3 \\ H_3N \quad\quad N_2 \\ NH_3 \end{array}\right]^{2+}$$

(i)

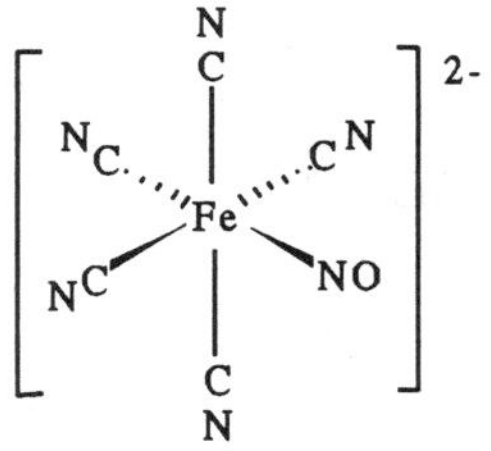

$$\begin{bmatrix} & N \\ & \| \\ & C \\ NC \cdots & | & \cdots CN \\ & Fe \\ NC & | & NO \\ & C \\ & \| \\ & N \end{bmatrix}^{2-}$$

(j)

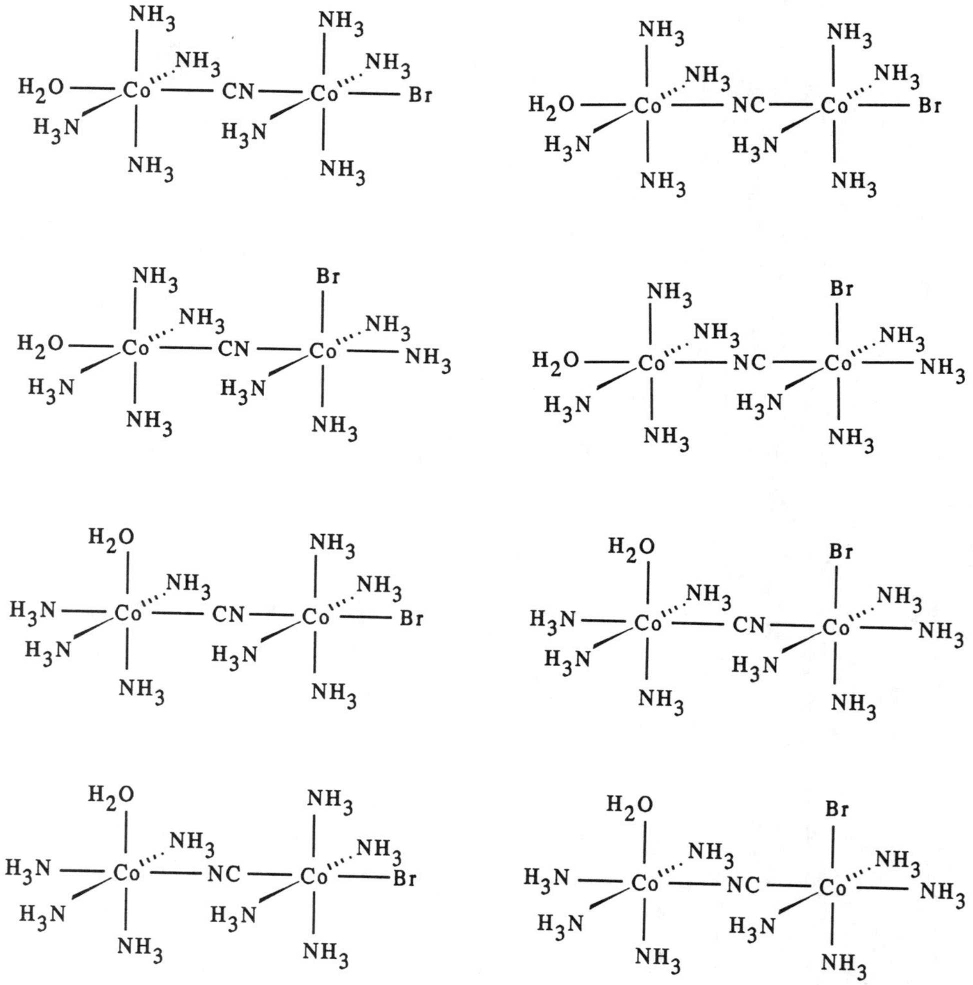

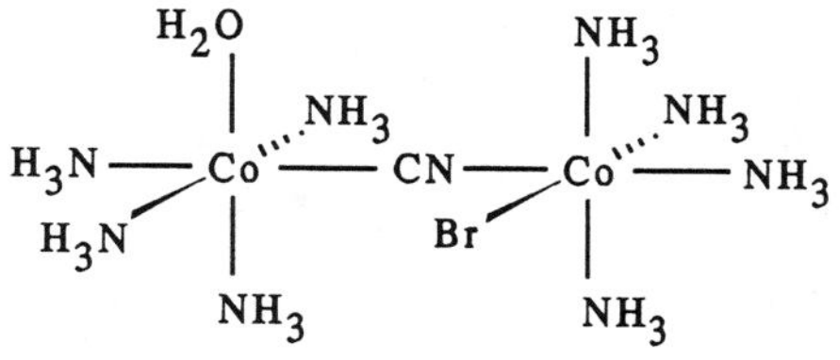

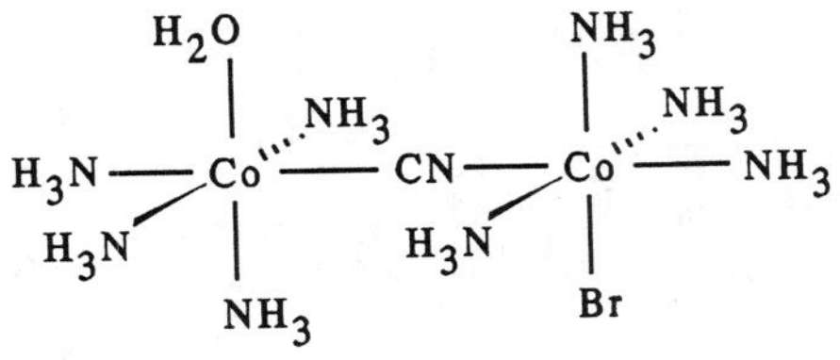

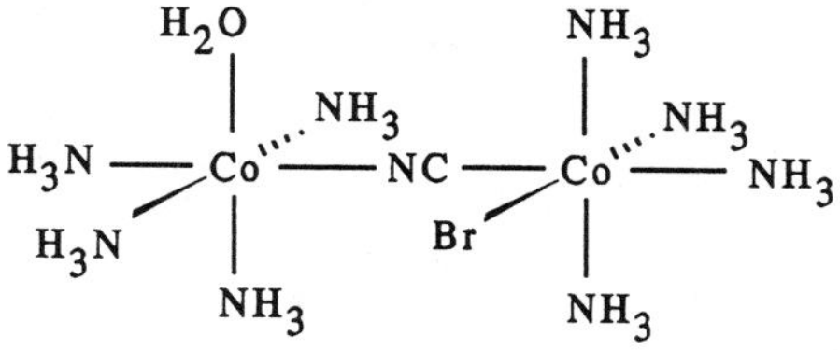

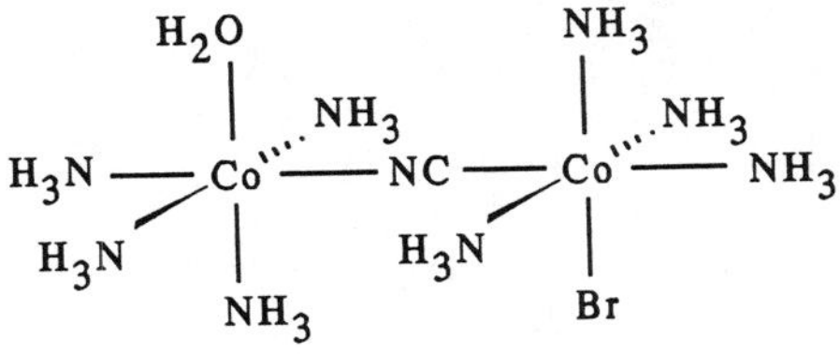

$$H_3N-\underset{\underset{NH_3}{|}}{\overset{\overset{H_2O}{|}}{Co}}\cdots NH_3 - CN - \underset{\underset{NH_3}{|}}{\overset{\overset{NH_3}{|}}{Co}}\cdots NH_3$$

(with H_3N, H_3N on left Co; Br below the CN bridge near right Co; NH_3 groups as drawn)

6-17B

As the charge density decreases in the series:

$$Li^+ > Na^+ > K^+ > Rb^+ > Cs^+$$

the rate of water exchange increases. This suggests that water ligands that are more tightly bound undergo substitution more slowly. This is characteristic of a dissociative (D or Id) mechanism.

6-19B

(a)

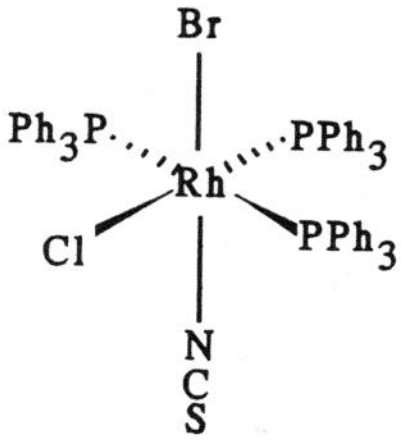

$$\underset{\underset{\underset{S}{\overset{C}{|}}}{\overset{N}{|}}}{\underset{Cl}{\overset{Ph_3P}{\underset{\cdots}{\overset{Br}{|}}}}}Rh\overset{PPh_3}{\underset{PPh_3}{\cdots}}$$

(b)

| trans
nitro | cis Δ
nitro | cis Λ
nitro |

| trans
nitrito | cis Δ
nitrito | cis Λ
nitrito |

6-21B

(a) See the mechanism of equations 6-5.25 - 6-5.27.

(b) See the mechanism of equation 6-5.23.

(c) Here we have induced aquation:

$$[Cr(NH_3)_5NCS]^{2+} + Hg^{2+} \rightarrow [Cr(NH_3)_5H_2O]^{3+} + HgNCS^+$$

6-23B

(a) Leaving group effects are important for substitution reactions of octahedral complexes, and linear free energy relationships have been observed, as in anation reactions. Octahedral complexes react predominantly by dissociative pathways (D or I_d mechanisms). For square complexes, leaving group effects are less important, although there is the question of the trans effect. Square complexes undergo substitution reactions by predominantly associative pathways (A or I_a mechanisms).

(b) Charge effects for substitution reactions of octahedral complexes are consistant with dissociative activation. Charge effects for square complexes are consistent with associative activation.

(c) Substitution reactions of octahedral complexes exhibit steric acceleration, as would be expected for dissociative activation. Steric hindrance would be expected for associative activation.

6-25B

(a) First we have attack of pentacoordinate $[Co(CN)_5]^{3-}$ at the sulfur atom of the thiocyanate ligand to give a ligand-bridged intermediate:

$$[*Co^{III}(NH_3)_5NCS]^{2+} + [Co^{II}(CN)_5]^{3-} \rightarrow [*Co^{III}(NH_3)_5-NCS-Co^{II}(CN)_5]^-$$

followed by inner sphere electron transfer and dissociation of the successor complex:

$$[*Co^{III}(NH_3)_5-NCS-Co^{II}(CN)_5]^- \rightarrow [*Co^{II}(NH_3)_5-NCS-Co^{III}(CN)_5]^-$$

$$[*Co^{II}(NH_3)_5-NCS-Co^{III}(CN)_5]^- \rightarrow$$
$$*Co^{2+}(aq) + 5NH_3 + [Co^{III}(CN)_5-SCN]^{3-}$$

(b) First we have formation of a ligand bridged binuclear intermediate:

$$[*Cr^{III}(H_2O)_5-SCN]^{2+} + [Cr^{2+}(aq)] \rightarrow [*Cr^{III}(H_2O)_5-SCN-Cr^{II}(H_2O)_5]^{4+}$$

followed by inner sphere electron transfer and dissociation of the successor complex:

$$[*Cr^{III}(H_2O)_5-SCN-Cr^{II}(H_2O)_5]^{4+} \rightarrow [*Cr^{II}(H_2O)_5-SCN-Cr^{III}(H_2O)_5]^{4+}$$

$$[*Cr^{II}(H_2O)_5-SCN-Cr^{III}(H_2O)_5]^{4+} \rightarrow [Cr(H_2O)_5-NCS]^{2+} + *Cr^{2+}(aq)$$

6-27B

(a) Water exchange rate constants for the aqua ions (Figure 6-6) depend inversely on charge density, for those ions lacking d electrons in the valence shell. First order rate constants for the rate determining step in anation reactions (k_O for reaction 6-5.12) show little dependence on the nature of the entering group. The values of k_O for anation approach the limit of water exchange rate constants, or in solvents other than water, approach the limits for solvent exchange in those solvents. Linear free energy relationships with slopes of 1 (or nearly so) have been discovered for certain acid hydrolysis reactions. Steric acceleration is typical for substitution reactions of octahedral complexes. Finally, charge effects are consistent with dissociation of the bond to the leaving group.

6-29B

The rate constant k_2 applies to the solvent-independent path in Figure 6-9. The data of this problem show an unmistakable entering group effect. This is typical of A or I_a mechanisms.

6-31B

This comparison indicates that aquation of a chloro ligand is more rapid for the complex with the smaller overall positive charge. A dissociative mechanism is indicated since breaking the bond to a leaving group (an anionic chloro ligand) is easier in the case of $[Co(en)_2Cl_2]^+$ than it is in $[Co(en)_2ClH_2O]^{2+}$.

6-33B

$$[Cr(H_2O)_5-NCS]^{2+} + Hg^{2+} \rightarrow [Cr(H_2O)_5-NCS-Hg]^{4+}$$
$$[Cr(H_2O)_5-NCS-Hg]^{4+} + H_2O \rightarrow [Cr(H_2O)_6]^{3+} + HgSCN^+$$

6-35B

At -22 °C and above, the molecule is fluxional. At -143 °C, the structure is:

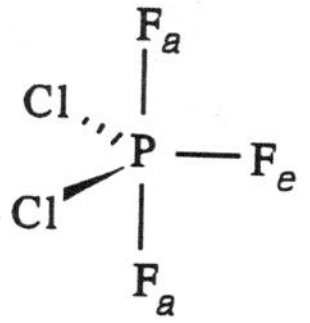

6-37B

In these stylized plots, the intercepts are the solvent-dependent rate constants, k_1 from the mechanism of Figure 6-9. The intercepts depend on the identity of the solvent. The slopes are the solvent-dependent rate constants, k_2 from the mechanism of Figure 6-9. The slopes depend on the identity of the entering ligand Y.

C - Questions from the Literature of Inorganic Chemistry

6-1C

(a) The authors discuss several points which seem to suggest that a D mechanism is unreasonable. First, the % incorporation of Cl^- ligand in the product mixture ($[Cr(H_2O)_6]^{3+}$ and $[Cr(H_2O)_5Cl]^{2+}$) is not the same for X = Br^-, I^- or SCN^-. A common, non-discriminating, five-coordinate intermediate (i.e. a D mechanism) should lead to equal Cl^- incorporation, regardless of the nature of the leaving group X. The authors point out that the % incorporation of Cl^- is also inconsistent with any known *cis* or *trans* effect series for the various ligands X = Br^-, I^-, NO_3^- or SCN^-. Finally, the behavior of the chromium complexes is shown to differ from that of similar cobalt ammines, which probably do react by dissociative pathways.

(b) The authors argue in favor of an I_a mechanism for the two reactions of pentaaquo chromium complexes:

$$[Cr(H_2O)_5X]^{2+} + H_2O \rightarrow [Cr(H_2O)_6]^{3+} + X^-$$
$$[Cr(H_2O)_5X]^{2+} + Cl^- \rightarrow [Cr(H_2O)_5Cl]^{2+} + X^-$$

and they argue in favor of an I_d mechanism for similar reactions of pentaammine complexes of Co^{III}:

$$[Co(NH_3)_5X]^{2+} + H_2O \rightarrow [Co(NH_3)_5H_2O]^{3+} + X^-$$
$$[Co(NH_3)_5X]^{2+} + Y^- \rightarrow [Co(NH_3)_5Y]^{2+} + X^-$$

(c) The authors cite the dependence upon the identity of X of the relative yields of $[Cr(H_2O)_6]^{3+}$ and $[Cr(H_2O)_5Cl]^{2+}$ (when $[Cr(H_2O)_5X]^{2+}$ is aquated in the presence of Cl^-) as evidence for assigning the I_a mechanism to the reactions of $[Cr(H_2O)_5X]^{2+}$. Conversely, aquation of $[Co(NH_3)_5X]^{2+}$ in the presence of anions such as Cl^- leads only to the aquo complex initially, suggesting that the ligand Cl^- is not able to compete with the entering ligand water, and making an I_d mechanism likely for such reactions of Co^{III} complexes.

(d) The I_a mechanism for the reactions of Cr^{III} complexes and I_d mechanism for the reactions of Co^{III} complexes are different from one another principally in the extent of interaction of entering or leaving groups with the first coordination sphere of the reactant complexes.

6-3C

(a) Three color changes were observed. The first color change results from formation of an ion pair (equation 1), as in the first step of an outer sphere electron transfer mechanism. This first step is "instantaneous." The second color change is due to a slower electron transfer (equation 2) to give $[Fe(CN)_6]^{3-}$ and a Co^{II} complex, which, being substitution labile, is rapidly sequestered by $EDTA^{4-}$ (equation 3). The $EDTA^{4-}$ was added to prevent precipitation of Co^{II} salts. The final color change is the redox reaction represented in the article by equation (4). This reaction is slow.
(b) Two consecutive outer sphere redox reactions take place: reactions (2) and (4). The second one is much slower than the first. The charge types (3+/4-) of the reactants in equations (1) and (2) are very high, and they favor rapid and strong ion pairing.

The reactant charge types (2-/3-) in reaction (4) do not favor ion pair interactions.

(c) Ion pair formation constants Q_{IP} have been calculated from experiment, and these are collected in Table I of the article. An electrostatic model (equations 6 and 7) has been used to estimate values of Q, assuming various distances a, the closest approach of the reactant ions. The agreement between Q_{IP} and Q is best if one uses the value of a that is estimated for "closest approach between the iron and the cobalt complexes on the side of the ammonia ligands."

A - Review

7-1A

The properties that determine the utility of a solvent are listed in section 7-1: temperature range of the liquid phase, dielectric constant (ε), Lewis acidity or basicity, Bronsted-Lowry acidity or basicity, and autoionization capacity.

7-3A

Solvents with good donor and/or acceptor abilities are well able to solvate ionic or polar solutes. The better the donor abilities of the solvent, the greater its ability to stabilize solute particles (especially cations) through donor-acceptor interactions. Polar molecular solvents interact with polar or ionic solutes to stabilize the solute. In the extreme application, we have solvents which are able to engage in Lewis acid-base interactions with solute particles.

7-5A

See structures 7-I through 7-III and equations 7-3.6 and 7-3.7.

7-7A

Among the aprotic solvents we have:

(a) nonpolar, non-ionizing molecular solvents which dissolve well only other such substances: hydrocarbons, etc.

(b) polar, non-ionizing molecular solvents which solvate polar or ionic substances generally through dipole-dipole interactions or through the formation of donor-acceptor adducts.

(c) polar and ionizing molecular solvents which have ionization equilibria such as those shown in equations 7-4.1 through 7-4.3.

7-9A

Solvents that are to be useful for electrochemical reactions must have a high dielectric constant and a wide range of electrochemical potentials over which they themselves are not oxidized or reduced.

7-11A

Bronsted-Lowry acids are proton donors, and bases are proton acceptors. Such acids serve to increase the concentration of the solvent's conjugate acid, $i.e.$ NH_4^+ in ammonia, H_3O^+ in water, or $H_3SO_4^+$ in sulfuric acid. The definition fails to describe non-protic systems.

7-13A

Acetic "acid" is acidic in water, due to the following equilibrium:

$$C_2H_3O_2H + H_2O = C_2H_3O_2^- + H_3O^+$$

In sulfuric acid solution, however, acetic "acid" serves as a base and increases the concentration of the solvent's "autoionization anion", HSO_4^-:

$$C_2H_3O_2H + H_2SO_4 = C_2H_3O_2H_2^+ + HSO_4^-$$

7-15A

By the Lewis definition, bases are electron pair donors and acids are electron pair acceptors. Examples are equations 7-8.18 and 7-8.19 as well as:

$$H^+ + CN^- = H-CN$$
$$H_3N: + BF_3 = H_3N{\rightarrow}BF_3$$
$$SnCl_2 + Cl^- = SnCl_3^-$$

7-17A

There is strong π-bonding in the planar BF_3 molecule. This is π-bonding of the AB_3 type shown in Chapter 3. The "electron deficiency" of this BX_3 system is well compensated by this intramolecular π-bond system, more so than in BBr_3. BBr_3 more readily accepts electrons from a σ donor, because its π-bond system is weaker than that in BF_3. Poor overlap between the boron 2p atomic orbital and the 4p atomic orbitals of the Br atoms is evidently responsible for the weakness of the π-bond system in BBr_3.

7-19A

This is equation 7-10.1.

7-21A

The strengths of the chlorine oxyacids follow the trend:
$HClO_4 > HClO_3 > HClO_2$.

The four remaining acids follow the trend in acidity: H_2SeO_4 (selenic acid) > H_2SeO_3 (selenous acid), by reason of the number of oxygen atoms. Arsenic acid, H_3AsO_4, is slightly weaker than phosphoric acid. $HMnO_4$ does not exist in aqueous solution, since the permanganate ion, MnO_4^-, readily decomposes in acidic solution giving MnO_2.

7-23A

Fluorosulfonic acid exhibits the autoionization process shown in reaction 7-13.2. The solvent autoionization cation is $H_2SO_3F^+$, and any substance that increases the concentration of this cation (beyond that typical of the pure solvent) is said to serve as an acid in the medium. SbF_5 serves in this fashion, as in reaction 7-13.3.

B - Additional Exercises

7-1B

The autoionization equilibrium for acetic acid is:

$$2C_2H_3O_2H = C_2H_3O_2H_2^+ + C_2H_3O_2^-$$

Acids are defined to be those substances which promote formation of the solvent cation:

$$H_2SO_4 + C_2H_3O_2H = C_2H_3O_2H_2^+ + HSO_4^-$$

$$HF + C_2H_3O_2H = C_2H_3O_2H_2^+ + F^-$$

Bases are defined to be those substances which promote formation of the solvent anion:

$$H_2O + C_2H_3O_2H = C_2H_3O_2^- + H_3O^+$$

$$NH_3 + C_2H_3O_2H = C_2H_3O_2^- + NH_4^+$$

Since its dielectric constant is lower than that of water, acetic acid will be a poorer solvent for ionic solutes.

7-3B
DMSO has a high dielectric constant and is a polar molecular substance that is well able to solvate both ions and polar solutes.

7-5B
The donor atom in the phosphines PR_3 is the somewhat "soft" phosphorus atom. The donor atom in the phosphine oxides R_3PO is the "hard" oxygen atom. The phosphine phosphorus atom can also enter into π-bonding with suitable Lewis acids.

7-7B
The hard, or class a metal ion Cr^{3+} should prefer the hard -NCS, *i.e.* the nitrogen-bound form. The soft, class b ion Pt^{II} should prefer the soft -SCN, *i.e.* the sulfur-bound form.

7-9B
$$2HCN = H_2CN^+ + CN^-$$

7-11B

Acids	Hybridization of the central atom	Adducts	Hybridization of the central atom
BF_3	sp^2	$F_3B\leftarrow L$	sp^3
$AlCl_3$	an ionic solid	$Cl_3Al\leftarrow L$	sp^3
$SnCl_2$	sp^2	$Cl_2Sn\leftarrow L$	sp^3

7-13B
This is an octahedral ion of the AB6 classification.

7-15B

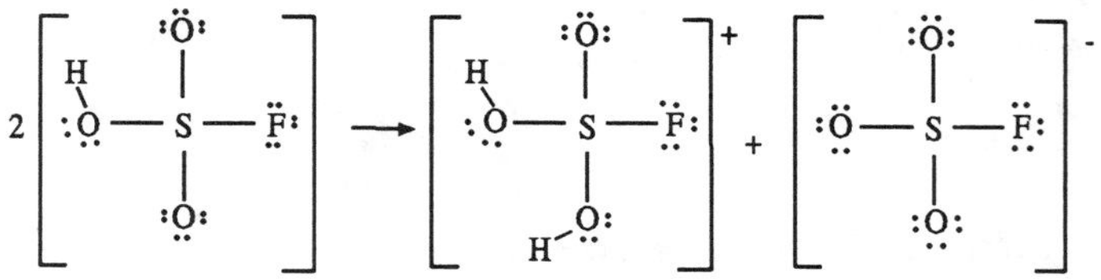

(a) The solvent system description develops as follows. Autoionization proceeds by proton transfer, as in the above equilibrium. Substances which serve in this solvent as acids will do so either by increasing the concentration of the cation $H_2SO_3F^+$ or by decreasing the concentration of the anion SO_3F^- compared to the values in the pure solvent. Correspondingly, substances which serve in this solvent as bases do so by enhancing the concentration of the solvent anion or by decreasing the concentration of the solvent cation (compared to the concentrations of these ions that are established through autoionization in the pure solvent).

(b) The Lewis approach to this equilibrium requires us to view the proton as the electron pair acceptor (a rather obvious view of the H+ ion). The Lewis base is the fluorosulfonic acid molecule that donates an electron pair to the proton, forming an O-H bond.

7-17B

The value of K_1 indicates that n (defined in section 7-12 to be the number of unprotonated oxygen atoms in an oxyacid) has the value 1 for each of these oxyacids of phosphorus. The structures are:

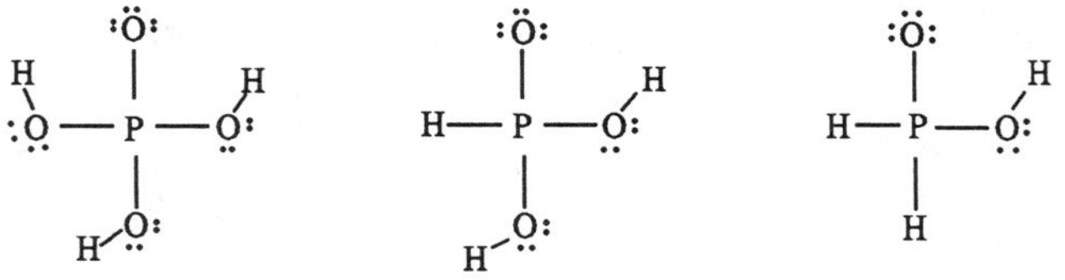

$$H_3PO_4 \qquad H_3PO_3 \qquad H_3PO_2$$

7-19B

(a) Bases are substances that dissolve and enter into ionization equilibria so as to increase the concentration of the solvent's autoionization anion beyond the value that is present in the pure solvent. Consider, for example, equation 7-8.8.

(b) Autoionization (also called self ionization) is the equilibrium in which either proton transfer, as for ammonia:

$$2NH_3 = NH_2^- + NH_4^+$$

or halide ion transfer, as for phosphorus pentachloride:

$$2PCl_5 = PCl_6^- + PCl_4^+$$

gives a solvent system autoionization cation and anion.

(c) Amphoterism is the ability to react either as an acid or as a base, as for $Al(OH)_3$ in the following:

$$Al(OH)_3 = Al^{3+} + 3OH^-$$

$$Al(OH)_3 = AlO_2^- + H_2O + H^+$$

6-21B

(a) SiO_2 is a Lux-Flood acid, and Na_2O is a Lux-Flood base.

(b) The hydride ion (H^-) is a Lewis base, and $B(OR)_3$ functions as a Lewis acid.

(c) N_2O_5 is the acid anhydride of HNO_3.

(d) The chloride ion functions as a Lewis base, and Cl_3PO functions as a Lewis acid.

(e) Lithium nitride is a base in liquid ammonia, because it produces the solvent's autoionization anion, NH_2^-.

(f) PF_5 dissolves in HF and produces an acidic solution, that is a solution with a higher concentration of the solvent's autoionization cation (H_2F^+) than exists in pure HF.

(g) PF_3 is a Lewis base, and $AlCl_3$ is a Lewis acid.

(h) BF_3 gives an acidic solution in ClF, since it produces the solvent's autoionization cation, Cl_2F^+. Alternately, ClF is a base in BF_3, giving the autoionization anion, BF_4^-.

(i) NOF is a base in ClF_3, since it produces the solvent's autoionization anion, ClF_4^-.

(j) XeO_3 is a Lux-Flood base (oxide donor), and $XeOF_4$ is a Lux-Flood acid (oxide acceptor).

(k) XeO_3 is a Lux-Flood acid.

(l) XeF_6 is a Lux-Flood acid. SiO_2 is a Lux-Flood base.

(m) ICl serves as an acid, giving the autoionization cation, PCl_4^+.

(n) Sulfur dissolves in ammonia, where it is seen to be an acid, since it gives the solvent's autoionization cation, NH_4^+.

7-23B

For $(CH_3)_3B:N(CH_3)_3$

$$5.79 \times 1.19 + 1.57 \times 11.20 = 24.47$$
$$\Delta H = -24.47 \text{ kcal/mol}$$

For $(CH_3)_3B:N(C_2H_5)_3$

$$5.79 \times 1.29 + 1.57 \times 10.83 = 24.47$$
$$\Delta H = -24.47 \text{ kcal/mol}$$

For $(CH_3)_3B:S(CH_3)_2$

$$5.79 \times 0.57 + 1.57 \times 6.49 = 13.49$$
$$\Delta H = -13.49 \text{ kcal/mol}$$

For $(CH_3)_3B:P(CH_3)_3$

$$5.79 \times 1.11 + 1.57 \times 6.51 = 16.65$$
$$\Delta H = -16.65 \text{ kcal/mol}$$

For $(CH_3)_3B:O(C_2H_5)_2$

$$5.79 \times 1.08 + 1.57 \times 3.08 = 11.09$$
$$\Delta H = -11.09 \text{ kcal/mol}$$

For $(CH_3)_3Al:N(CH_3)_3$

$$17.32 \times 1.19 + 0.94 \times 11.20 = 31.14$$
$$\Delta H = -31.14 \text{ kcal/mol}$$

For $(CH_3)_3Al:N(C_2H_5)_3$

$$17.32 \times 1.29 + 0.94 \times 10.83 = 32.52$$
$$\Delta H = -32.52 \text{ kcal/mol}$$

For $(CH_3)_3Al:S(CH_3)_2$

$$17.32 \times 0.57 + 0.94 \times 6.49 = 15.97$$
$$\Delta H = -15.97 \text{ kcal/mol}$$

For $(CH_3)_3Al:P(CH_3)_3$

$$17.32 \times 1.11 + 0.94 \times 6.51 = 25.35$$
$$\Delta H = -25.35 \text{ kcal/mol}$$

For $(CH_3)_3Al:O(C_2H_5)_2$

$$17.32 \times 1.08 + 0.94 \times 3.08 = 21.61$$
$$\Delta H = -21.61 \text{ kcal/mol}$$

For $(CH_3)_3Ga:N(CH_3)_3$

$$13.83 \times 1.19 + 0.40 \times 11.20 = 20.94$$
$$\Delta H = -20.94 \text{ kcal/mol}$$

For $(CH_3)_3Ga:N(C_2H_5)_3$

$$13.83 \times 1.29 + 0.40 \times 10.83 = 22.17$$
$$\Delta = -22.17 \text{ kcal/mol}$$

For $(CH_3)_3Ga:S(CH_3)_2$

$$13.83 \times 0.57 + 0.40 \times 6.49 = 10.48$$
$$\Delta H = -10.48 \text{ kcal/mol}$$

For $(CH_3)_3Ga:P(CH_3)_3$

$$13.83 \times 1.11 + 0.40 \times 6.51 = 17.95$$
$$\Delta H = -17.95 \text{ kcal/mol}$$

For $(CH_3)_3Ga:O(C_2H_5)_2$

$$13.83 \times 1.08 + 0.40 \times 3.08 = 16.17$$
$$\Delta H = -16.17 \text{ kcal/mol}$$

CHAPTER 8

A - REVIEW

8-1A

Among the elements we have,

(a) gases: N_2, O_2, the halogens and the noble gases.

(b) liquids: Ga (mp = 30 °C), and Hg (mp = -39 °C).

(c) solids which melt below 100 °C: Na, K, Rb, Cs, Ga, P.

8-3A

See the S_8 molecule of Figure 8-1.

8-5A

The electronic structures are written in Figure 2-12.

(a) The ionization enthalpy for Li (527 kJ mol^{-1}) is lower than those of any other element in row two, because the outer-most electron of Li ($1s^2 2s^1$) is the first electron added to the shell with principal quantum number n = 2, and it is therefore easily removed. Low first ionization enthalpies and ready reactivities with oxidizing agents are characteristic of all the metals in Group IA(1).

(b) One might expect the 2+ ion, since Be is the first member of Group IIA(2). The ionization enthalpies of Be, however, are very high, and the normal chemistry of the element is that of its covalent substances. The charge density in the aqueous Be^{2+} ion is so high that hydrolysis results, giving acidic solutions containing species with Be-OH bonds.

(c) The discontinuity is illustrated in Figure 8-11. The half-filled, triply degenerate set of 2p orbitals of the nitrogen atom is disrupted upon making the configuration ($1s^2 2s^2 2p^4$) of oxygen. The fourth of these 2p electrons for oxygen must be paired with one of the other three. There is a loss in stability (decrease in ionization enthalpy) due to this pairing and due also to a decrease in the number of unpaired electrons compared with nitrogen.

(d) There is a general trend for electron attachment enthalpies to become more negative from left to right in a row of the periodic

table. It may not be wise to say more than this. The periodic trend
should in some way parallel that of electronegativities, the elements
to the right in any row being the ones more likely to form anions.

(e) See the answer to 5d above.

8-7A

The octet rule is discussed in the early portions of Chapter 3. It
is the principle that guides us in drawing Lewis diagrams, namely
that an element in a covalent substance typically acquires eight
electrons, either lone electrons or shared, bonding electrons. The
rule quite obviously does not apply to hydrogen, nor does it apply to
elements of rows three and below, which are able to acquire more than
a total of eight electrons in forming reasonable substances.

8-9A

Whereas silicon has the diamond structure (as do the other elements
of Group IVB(14)), it does not adopt the graphite structure. This is
because, of the elements in Group IVB(14), only carbon successfully
enters into the sort of $p\pi$-$p\pi$ bonding that characterizes the layers
of graphite (Figure 8-3 and structure 8-III).

8-11A

The elements of Group IIA(2) are Be, Mg, Ca, Sr, Ba, and Ra. The
elements of Group IIB(12) are Zn, Cd, and Hg. Although the elements
of each group characteristically form divalent cations, those of Zn
and Cd are much more polarizable because of the filled $3d^{10}$
configuration. This means that Zn and Cd form complexes with ligands
such as NH_3 and X^- much more readily than do Mg or Ca, in spite of
the fact that the radii of these cations are similar. Mercury is
unique in Group IIB(12) because of the high, positive reduction
potential of the divalent cation and because of the stable metal-
metal bond of the dimeric mercurous (Hg^I) ion, Hg_2^{2+}. Beryllium is
unique among the elements of Group IIA(2) because its first and
second ionization enthalpies are so high. Even the most oxidizing
elements do not react with elemental beryllium to give compounds with
completely ionic character. Thus BeF_2 has high covalent character in

the Be-F bonds. Also, the aqueous Be^{2+} ion is greatly hydrolyzed, giving acidic solutions.

8-13A

The noble gas configuration for N is attained through formation of compounds having occupancy formulas (AB_xE_y, as defined in Chapter 3) of the following types:

 (a) ABE, for example N_2 and RCN.

 (b) AB_2E, for example NO_2^- and N_2H_2.

 (c) AB_3E, for example NH_3 and N_2H_4.

 (d) AB_4, for example NR_4^+.

 (e) AB_3, for example NO_3^-.

See also Chapter 16.

Sulfur achieves the noble gas configuration in a greater variety of ways, owing to its position in row three, and to the consequent availability of d orbitals for bonding. We discuss the variety of compounds formed by sulfur in Chapter 19. For now we should note the following: S^{2-} and HS^-, H_2S, R_3S^+, SO_2, SO_3, SO_3^{2-}, SO_4^{2-}, SF_5^-, SF_6, etc.

8-15A

It is typical of the transition metals that they:

 (a) are hard, high-melting metals.

 (b) are good conductors of heat and electricity.

 (c) are easily alloyed with other metals.

 (d) are variously reactive with one or another of the mineral acids.

 (e) exhibit a variety of valences, oxidation states, colors and magnetisms.

It is also typical of the transition metals and their ions that they form a variety of inorganic and organometallic coordination compounds. See Chapters 23-29 especially.

8-17A

The icosahedron is the main structural feature of the element boron.

CHAPTER 8

8-19A

The chemistry of hydrogen involves one of three electronic "processes": ionization to H^+, reduction to H^-, or formation of a covalent bond.

8-21A

The lanthanide contraction is the steady decrease in atomic size that takes place starting with lanthanum and proceeding through the series. It is caused by the progressively imperfect shielding of one f orbital by the others, such that the effective nuclear charge increases steadily across the lanthanum row. The major consequence of this increasing effective nuclear charge is that elements that follow the lanthanides are smaller than might be expected solely on the basis of atomic number.

B - Additional Exercises

8-1B

These diatomic elements were discussed in Chapter 3. The appropriate energy level diagrams are given in Figure 3-26. The MO electron configurations are:

$$N_2: \quad [\sigma_1]^2[\sigma_2]^2[\pi_1]^4[\sigma_3]^2$$

$$O_2: \quad [\sigma_1]^2[\sigma_2]^2[\sigma_3]^2[\pi_1]^4[\pi_2]^2$$

$$F_2: \quad [\sigma_1]^2[\sigma_2]^2[\sigma_3]^2[\pi_1]^4[\pi_2]^4$$

Oxygen is paramagnetic because of the two unpaired electrons that are placed into the doubly degenerate π_2 level. This set of π molecular orbitals is filled for F_2, which is diamagnetic. The bond orders are F_2, one; O_2, two; and N_2, three.

8-3B

"BH_3" exists as the dimer, B_2H_6, diborane(6), discussed in Chapters 3 and 12. In a formal sense, though, we might consider that "BH_3" is a Lewis acid because it "forms" the adducts $L{\rightarrow}BH_3$, where L represents a Lewis base. The B-H bonds in diborane are largely covalent, and the substance does not tend to ionize.

HF on the other hand is a covalent, gaseous substance that ionizes readily in water to give H^+, as in equation 8-7.1. The H-F bond is a polar covalent one, and the polarity in the Lewis diagram can be represented as follows:

$$\delta^+ \quad H-F \quad \delta^-$$

CH_4 is a covalent, nonpolar molecular hydride with tetrahedral geometry, since the carbon atom is of the AB_4 classification discussed in Chapter 3. Methane does not behave as a proton donor, a proton acceptor, Lewis acid or Lewis base in its normal solution chemistry. Methane is not hydridic.

NH_3 is a pyramidal molecule, and the nitrogen atom is of the AB_3E classification, having an occupancy (defined in Chapter 3) of four. The molecule is a Lewis base by reason of the lone pair on nitrogen, and it behaves as a Bronsted base, accepting a proton to give NH_4^+. Ammonia may ionize to give NH_2^-, but powerful reducing agents are required, e.g. Na, such that ammonia is otherwise best considered to be basic. Recall the material of Chapter 7, and especially the fact that the autoionization constant for NH_3 is much less than that of water.

As does the nitrogen atom of ammonia, the oxygen atom of water has an occupancy of four. The oxygen is of the AB_2E_2 type, however, and there are two lone pairs on oxygen. Water can act as a Lewis base, a Bronsted base and a Bronsted acid. The O-H bonds are of the polar covalent type.

HF, H_2O and NH_3 are capable of intermolecular hydrogen-bonding.

8-5B

		pπ Donor Atom(orbital)	dπ Acceptor Atom(acceptor)
(a)	Cl_3P	$O(2p)$	$P(3d)$
(b)	Cl_2SO, SO_2, SO_4^{2-}	$O(2p)$	$S(3d)$
(c)	ClO_2, ClO_4^-	$O(2p)$	$Cl(3d)$
(d)	PO_4^{3-}	$O(2p)$	$P(3d)$

In each case the pπ→dπ donation takes place *via* the sort of overlap depicted in Figure 3-14. This overlap may involve more than a single oxygen atom donor, such that the resulting pπ→dπ interaction is delocalized over, in the case of PO_4^{3-} for instance, five atoms.

8-7B

The f orbitals lie so close to one another in energy that the distinction between one configuration and another is not great. Also, the f-orbitals do not extend to the periphery of the various lanthanide atoms and ions, so that the f-electrons are not much influenced by the presence of different numbers or varieties of surrounding groups.

8-9B

See Section 13-9.

8-11B

See Section 17-11.

A - Review

9-1A

The three isotopes of hydrogen (and their abundances) are ^{1}H, hydrogen; ^{2}H, deuterium (0.0156 %); and ^{3}H, tritium (ca. 1 part in 10^{17}). Only the latter is radioactive.

9-3A

The H-H bond strength is high, as indicated by the high positive value of ΔH° for reaction 9-1.6. It is the strength of the homonuclear hydrogen bond that makes hydrogen so unreactive.

9-5A

The familiar representation of the hydrogen bond (X-H...Y) involves a short, covalent bond to hydrogen (X-H) and a longer interaction (H...Y) between hydrogen and an electron-rich atom Y.

9-7A

Hydrogen bond enthalpies are typically in the range 20-30 kJ mol^{-1}.

9-9A

The water in crystalline hydrates is usually bound into the lattice in one of the three ways shown in Figure 9-3.

9-11A

The chlorine hydrate, $Cl_2 \cdot 7.3H_2O$, is really a clathrate. The compound, formed by the cystallization of water vapor in the presence of chlorine, involves incomplete filling of the various cavities in the clathrate structure.

9-13A

Among the binary hydrides, there are first the saline hydrides of the Group IA(1) metals and the following Group IIA(2) elements: CaH_2, SrH_2, and BaH_2. These are ionic substances that contain the hydride anion, H$^-$. Among these, LiH is substantially more covalent than the others. BeH_2 is a hydrogen-bridged covalent polymer. MgH_2 falls on the borderline between ionic and covalent character.

The hydrides of the elements from Groups IIIB(13) through VIIB(17)
are covalent and have varying properties. The halogen hydrides, HX,
are acidic. The hydrides of N and P are basic. H_2O is amphoteric.
As a class, the covalent hydrides are mostly molecular in nature, as
summarized in the text.

The transition metal hydrides are either ionic compounds of a
stoichiometric nature (usually MH_2 or MH_3) or interstitial,
nonstoichiometric hydrides, some typical examples being $LaH_{2.87}$,
$YbH_{2.55}$, and $TiH_{1.7}$.

9-15A

(a) A hydrated substance contains a fixed and definite number of
water molecules, which occupy specific positions in the crystal
structure of the solid. Examples are $ScCl_3 \cdot 6H_2O$ and $CoCl_2 \cdot 6H_2O$.

(b) A hydrous compound contains water in varying proportions.
Examples are the noncrystalline hydrous oxides such as colloidal iron
oxide, $Fe_2O_3 \cdot xH_2O$ (Chapter 5).

(c) A gas hydrate is a clathrate of a gas. The gas is trapped in the
open regions of a crystalline form of water. Gas hydrates are formed
by co-condensation of a gas such as Cl_2 or SO_2 with water. The water
host crystallizes with an open structure, and the gas molecules
reside in cavities in the structure. An example is the chlorine
hydrate, $Cl_2 \cdot 7.30H_2O$.

(d) A liquid hydrate is formed when liquids such as chloroform
crystallize with water to form a clathrate compound. The liquid is
found trapped in cages formed by the water molecules.

(e) Salt hydrates are clathrate compounds formed when salts
crystallize with high water content. They are usually sulfonium or
R_4N^+ salts such as tetrabutylammonium benzoate, which crystallizes as
$[(C_4H_9)_4N][C_6H_5CO_2] \cdot 37.5H_2O$.

B - Additional Exercises

9-1B

A saline hydride such as LiH or NaH characteristically reacts with a Bronsted acid to give hydrogen. Two preparations of HD are suggested. The first is reaction of a hydride with a deuterated acid:

$$LiH + D_2O \rightarrow HD + LiOD$$

The second is reaction of a deuteride with a protic acid:

$$NaD + H_2O \rightarrow HD + NaOH$$

9-3B

The principal factor lending strength to a hydrogen bond seems to be electrostatic. Other effects may contribute to the interaction, and the situation changes dramatically from one H-bond to the next. Nevertheless, it is the Coulomb interaction that dominates the strength of most H-bonds, so we expect the S-H...O interaction to be stronger than the O-H...S one, since oxygen has the greater negative character.

9-5B

(a) $CaH_2 + 2H_2O \rightarrow 2H_2 + Ca(OH)_2$

(b) $B_2H_6 + 2NaH \rightarrow 2NaBH_4$

(c) $2SiCl_4 + 2LiAlH_4 \rightarrow 2SiH_4 + 2LiCl + Al_2Cl_6$

(d) $Al_2Cl_6 + 8LiH \rightarrow 2LiAlH_4 + 6LiCl$

9-7B

Water reacts with hydridic (H^-) substances to form molecular hydrogen:

$$NaH + H_2O \rightarrow H_2 + NaOH$$

Water reacts with covalent ($H\cdot$) hydrogen-containing substances as in steam re-forming of methane:

$$CH_4 + H_2O \rightarrow CO + 3H_2$$

Water reacts with acidic substances (H^+) to give the hydronium ion:

$$HCl + H_2O \rightarrow H_3O^+ + Cl^-$$

9-9B

Treatment of aqueous solutions of the anions gives H_2S and H_2Se:

$$Na_2S + 2HCl \rightarrow H_2S + 2NaCl$$
$$K_2Se + 2HCl \rightarrow H_2Se + 2KCl$$

Saline hydrides are used as reducing agents for the preparations of $NaBH_4$ and $LiAlH_4$:

$$B_2H_6 + 2NaH \rightarrow 2NaBH_4$$
$$Al_2Cl_6 + 8LiH \rightarrow 2LiAlH_4 + 6LiCl$$

Saline hydrides may be prepared by direct reaction of the elements:

$$Na + 1/2H_2 \rightarrow NaH$$
$$Ca + H_2 \rightarrow CaH_2$$

9-11B

(a) $CaH_2 + 2H_2O \rightarrow Ca(OH)_2 + 2H_2$

(b) $K + C_2H_5OH \rightarrow KOC_2H_5 + 1/2H_2$

(c) $KH + C_2H_5OH \rightarrow KOC_2H_5 + H_2$

(d) $UH_3 + 2H_2O \rightarrow UO_2 + 7/2H_2$

(e) $UH_3 + 2H_2S \rightarrow US_2 + 7/2H_2$

(f) $UH_3 + 3HCl \rightarrow UCl_3 + 3H_2$

9-13B

(a) $2Na + H_2 \rightarrow 2NaH$

(b) $H_2 + B_2H_6 \rightarrow$ higher boranes (See Chapter 12.)

(c) $Ca + H_2 \rightarrow CaH_2$

(d) $2Li + H_2 \rightarrow 2LiH$

(e) $N_2 + 3H_2 \rightarrow 2NH_3$

(f) $O_2 + 2H_2 \rightarrow 2H_2O$

(g) $U + 3/2H_2 \rightarrow UH_3$

9-15B

This is the same unit cell as for NaCl, except that H^- replaces Cl^- in Figure 4-1. The coordination number of sodium in NaH is the same as that in NaCl, namely six.

9-17B

$$LiGaH_4 \rightarrow Li + Ga + 3/2H_2$$

The adduct of AlH_3 with H^- is more stable than its gallium counterpart. The M-H force constants and bond strengths in AlH_4^- are higher than those in GaH_4^-. There is a greater M-H bond covalency in lithium aluminum hydride.

CHAPTER 10

A - Review

10-1A

The alkali metals are soft and volatile because the metallic bond for
these metals is weak. There is a single s-electron in the valence
shell for each metal, and when the metals adopt close-packed solid
structures, the binding energy is weak. Refer to the discussion of
the metallic bond in section 8-6 and to Figures 8-8, 8-9, and 8-10.
Because the metallic bond is weak, the inter-atomic forces are
overcome easily, and the metals have low vaporization enthalpies, as
shown in Figure 8-10. Mercury, having a similarly weak metallic
bond, is a liquid.

10-3A

We can write this simply: $[Rn]7s^1$. Note that this is an electron
configuration which includes a filled 4f shell and filled 3d, 4d, and
5d shells.

10-5A

The size of the atom increases, and the ionization enthalpy (Figure
2-14) decreases slightly, from top to bottom in the group. The
metallic bond is weaker for the elements near the bottom of the
group, as indicated by the decrease from top to bottom in the group,
in values for ΔH_{vap} (Figure 8-10). All of these factors indicate
that the reactivity of the metal (i.e. disruption of the metallic
bond and ionization to give a cation) should increase from top to
bottom in the group.

10-7A

Li^+ has a very high charge-to-radius ratio, and its chemistry is
similar to that of the dipositive ion Mg^{2+}. Lithium is the only
alkali metal that reacts directly with nitrogen to give a stable
nitride, LiN_3. Also, reaction of lithium with molecular oxygen leads
to an oxide, Li_2O, presumably because the small cation favors an
oxide (small anion) lattice. The other alkali metals give peroxides
or superoxides upon reaction with molecular oxygen. An additional
consequence of the small size of Li^+ is that the hydroxide is less

soluble than the hydroxides of other alkali metals. LiOH, unlike
other alkali metal hydroxides, is unstable at high temperatures to
give the oxide, Li_2O.

10-9A

Dilute solutions of the alkali metals in liquid ammonia contain
solvated metal cations and solvated electrons. More concentrated
solutions contain metal atom clusters and have conductivities typical
of metallic substances. Ammonia solutions of the alkali metals are
powerful reducing agents.

10-11A

The structures of NaCl and CsCl are given in Figure 4-1. The
structures are different because the Cs^+ cation is larger than the
Na^+ cation. Recall from Chapter 4 that the coordination number of
cesium in CsCl is 8, whereas the coordination number of sodium in
NaCl is 6. The sodium cation is too small to allow NaCl to adopt the
CsCl structure, as shown by the data of Table 4-3.

10-13A

The preference of the anionic sites on the resin for the alkali
cations (A^+(aq) in equation 10-7.1) follows the order $Cs^+ > Rb^+ > K^+$
$> Na^+ > Li^+$. The larger Cs^+ ion is less hydrated and more able to
approach the anionic binding sites on the resin.

10-15A

A crown ether is a simple cyclic polyether such as that shown in
Structure 10-I. A cryptate is polycyclic and has donor atoms other
than oxygen, an example being Structure 10-VI. A cryptate offers
advantages over a crown ether in complexing a metal cation because it
is polycyclic, and can surround the cation more completely. The
donor atoms other than oxygen also increase the cryptate's complexing
abilities.

10-17A

Sodium fires are best extinguished with sand or inert powders. Some
common chemical fire extinguishers contain substances (CO_2, H_2O or
CCl_4) that would react with the sodium.

B - Additional Exercises

10-1B

The diatomic molecules of the alkali metals (M_2) have the general MO
electron configuration $[\sigma_{ns}]^2$. Dilithium, for example, has the MO
configuration $[\sigma_{2s}]^2$, shown in Figure 3-26, and the electron
distribution shown in Figure 3-25. The bond order is 1.0, as for the
other M_2 molecules. From top to bottom in Group IA(1), the
ionization enthalpy (ΔH_{ion}) decreases, and the valence ns orbital
increases in energy. The ns orbitals which overlap in forming the σ
MO become increasingly diffuse as their energies increase from top to
bottom in the group.

10-3B

We estimate that the "hole" created within the crown is well-matched
for the K^+ cation, whose radius is 1.33 Å. Metal ions with radii
either too small or too large for this cavity give lower stability
constants.

10-5B

The cryptate ligands generally form more stable complexes with simple
cations than do the crown ethers because the former more effectively
surround the cation and because the former have donor atoms such as
nitrogen as well as oxygen. The complexation of metal cations by
cryptates is so favorable that normally insoluble ionic substances
such as $BaSO_4$ may be dissolved in the presence of the cryptate
ligands.

10-7B

(a) $KCl(l) + Na(g) \rightarrow NaCl(l) + K(g)$

(b) $3Li + 1/2N_2 \rightarrow Li_3N$

(c) $2Na + O_2 \rightarrow Na_2O_2$

(d) $Cs + O_2 \rightarrow CsO_2$

(e) $K + C_2H_5OH \rightarrow K^+C_2H_5O^- + 1/2H_2$

(f) $Li(s) + Et_2NH(l) \rightarrow LiNEt_2(s) + 1/2H_2$

(g) $Li + NH(SiMe_3)_2 \rightarrow LiN(SiMe_3)_2 + 1/2H_2$

(h) $2RbO_2 + 2H_2O \rightarrow O_2 + 2Rb^+ + 2OH^- + H_2O_2$

(i) $Li_2O + H_2O \rightarrow 2Li^+ + 2OH^-$

(j) $2KOH + CO_2 \rightarrow 2K^+ + CO_3{}^{2-} + H_2O$

(k) $K^+(aq) + B(C_6H_5)_4{}^-(aq) \rightarrow K[B(C_6H_5)_4](s)$

(l) $C_6H_5Cl + 2Li \rightarrow LiC_6H_5 + LiCl$

(m) $n\text{-}C_4H_9Li + CH_3I \rightarrow LiCH_3 + n\text{-}C_4H_9I$

(n) $LiCH_3 + [W(CO)_5Cl]^- \rightarrow [W(CO)_5CH_3]^- + LiCl$

10-9B

The high charge-to-radius ration of Li causes its compounds to have a
more pronounced covalency. Of the alkali metals, only Li is reactive
towards the acidic hydrogen of $C_6H_5C \equiv CH$. Also, Li reacts only
slowly with water. Only Li reacts with N_2 to give a nitride
directly. With oxygen, Li gives mostly Li_2O, with traces of Li_2O_2.
LiH is more stable thermally than the saline hydrides of the other
Group IA(1) metals. LiOH decomposes at high temperatures to give
Li_2O, but the hydroxides of the other Group IA(1) metals only
sublime. Similarly, Li_2CO_3 is thermally unstable and gives Li_2O plus
CO_2. The solubilities of lithium salts more closely resemble those
of Mg^{2+}. Lithium halides are even soluble in certain organic
solvents such as acetone and ethyl acetate. Lithium salts are more
commonly hydrated, cf. $LiClO_4 \cdot 3H_2O$ and $NaClO_4$. Lithium ion also has
larger hydration numbers and energies, as shown in Table 10-1.
Organo lithium compounds are highly covalent in nature, and they are
both soluble in nonpolar organic solvents and volatile. The alkyls
of Li are strongly aggregated in the solid and in solution, as for
the tetramer shown in Figure 10-4. These differences can be
attributed to the small size of the lithium atom.

10-11B

The Li^+ cation is chelated by the crown ether, disrupting the
aggregates between Li^+ and the methyl anion. This releases the
methyl groups, which are found to react more readily.

10-13B

(a) The primary coordination number is only four, but the ion in solution is also bound to as many as 20 additional water molecules, as indicated by the data of Table 10-1.

(b) See Figure 10-1.
(c) This is similar to Structure 10-V.
(d) See Figure 10-4.

10-15B

The unit cell edge is 4.08 Å, and the LiH distance is thus half this value, 2.04 Å. By Pauling's method, we first calculate the effective nuclear charges of these isoelectronic ions:

$$Z^*(H^-) = 1 - 0.35 = 0.65$$
$$Z^*(Li^+) = 3 - 0.35 = 2.65$$

The ratio of the ionic radii is thus:

$$\frac{r_{Li^+}}{r_{H^-}} = \frac{Z^*_{H^-}}{Z^*_{Li^+}} = \frac{0.65}{2.65} = 0.25$$

Then $r_+ = 0.25 \times r_-$. Solving for r_-, we get:

$$2.04 \text{ Å} = 1.25 \times r_-$$
$$r_- = 1.63 \text{ Å}$$

The radius of the cation is then: $r_+ = 2.04 - 1.63 = 0.41$ Å.

10-17B

$$2Na(s) + O_2(g) \rightarrow 2Na(g) + O_2(g) \qquad 2\Delta H_{vap}$$

$$2Na(g) + O_2(g) \rightarrow 2Na(g) + 2O(g) \qquad \Delta H_{diss}$$

$$2Na(g) + 2O(g) \rightarrow 2Na^+(g) + 2O(g) \qquad 2\Delta H_{ion}$$

$$2Na^+(g) + 2O(g) \rightarrow 2Na^+(g) + 2O^{2-}(g) \qquad 2\Delta H_{EA}(1) + 2\Delta H_{EA}(2)$$

$$2Na^+(g) + 2O^{2-}(g) \rightarrow Na_2O_2(s) \qquad \Delta H = U$$

$$2Na(s) + O_2(g) \rightarrow Na_2O_2(s), \Delta H = 2\Delta H_{vap} + \Delta H_{diss} + 2\Delta H_{ion}$$
$$+ 2\Delta H_{EA}(1) + 2\Delta H_{EA}(2) + U$$

C - Questions from the Literature of Inorganic Chemistry

10-1C

(a) Some of the ligands are shown in Figure 1 of the paper. Dicyclohexyl-18-crown-6 is shown in the text, structure 10-III. The remainder have the following structures:

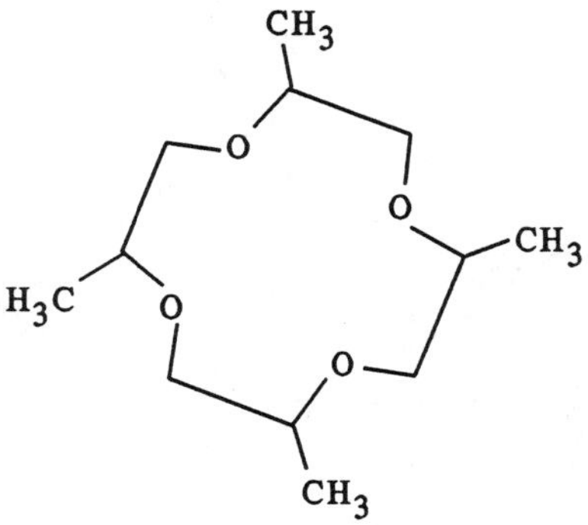

tetramethyl-12-crown-4

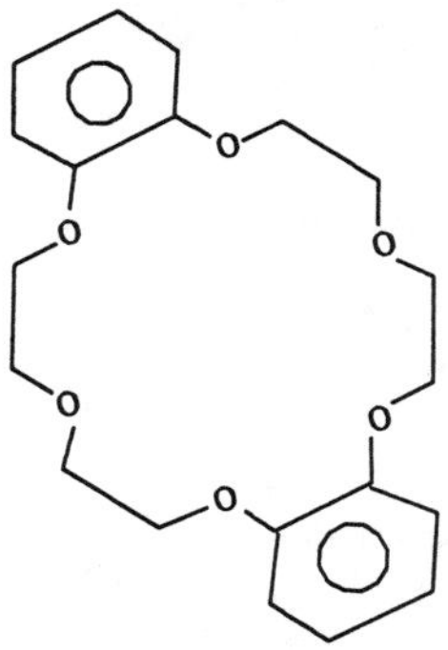

dibenzo-18-crown-6

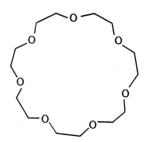

21-crown-7

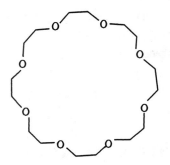

24-crown-8

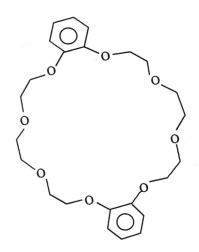

dibenzo-24-crown-8

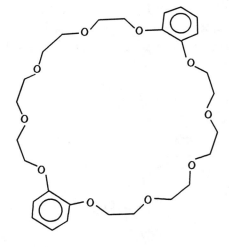

dibenzo-30-crown-10

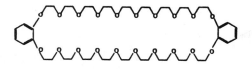

dibenzo-60-crown-20

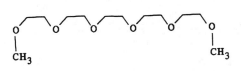

pentaglyme

(b) The stability constants of the Na^+, K^+ and Cs^+ complexes with various crown ethers in ethanol are given in Table II, and a comparison of ion and crown-hole sizes is given in Table III. The large Cs^+ cation (3.34 Å diameter) favors crowns with the larger ring sizes, i.e. 21-crown-7. The optimum ring size for K^+ (2.66 Å cation diameter) is 18, as in 18-crown-6. Na^+ complexation is favored over potassium if the crown ring size is 15 to 18, i.e. dicyclohexyl-14-crown-4.

(c) The macrocyclic nonactin and valinomycin cryptands are able to surround K^+ by wrapping around the cation completely. This precludes any further interaction of the cation with the solvent. This is not the case for most of the complexes of K^+ with smaller crown ethers. In the latter cases, some interaction between K^+ and solvent is "possible in the direction perpendicular to the plane of the ring." For the larger crown ethers however (*viz* dibenzo-30-crown-10 of Table II), extensive wrapping around the potassium cation may account for the extra stability of these complexes.

(d) Stability constants measured in water are lower than those measured in methanol. This may be due to a greater solvation of the free cation by water than by methanol. Such a difference in solvation of the free cation would cause the solvated cation to be more favored over the crown-complex in water than in methanol.

10-3C

(a) These derivatives are very much like the simple alkyl amides that are found in equations 10-3.4 and 10-3.5 of the text. The two substituents on nitrogen are trimethylsilyl groups, $-SiMe_3$. We can write the Lewis diagram for this bis(trimethylsilyl) amide anion as follows:

The group is bulky and there is π-delocalization over the N and the two Si atoms. The large size of the anion disposes it to form three-coordinate complexes, because there is not room for more than three of these ligands at any one metal center. π-bonding between the metals and these amide ligands is discussed in the paper.

(b) These salts were prepared in a manner that is analogous to reactions 10-3.4 and 10-3.5: reaction of a bis(trimethylsilyl)amine with the strong base n-butyl-lithium, $CH_3CH_2CH_2CH_2^-Li^+$. The butyl anion deprotonates the secondary amine, leaving an amide.

(c) The three-coordinate complexes of Table 2 have trigonal geometries.

(d) The general reaction used in these syntheses is:

$$MCl_3 + 3Li[N(SiMe_3)_2] \rightarrow M[N(SiMe_3)_2]_3 + 3LiCl$$

(e) The bis(trimethylsilyl)amide anion has a $d\pi$-$p\pi$ bond system delocalized over the N atom and the two Si atoms:

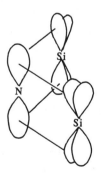

This anion is furthermore able to engage a metal in π
bonding, either through metal-to-N π donation:

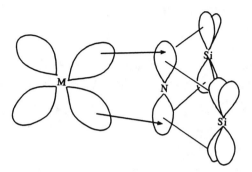

or through N-to-metal π donation:

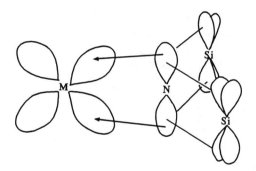

In the first of these two π bonds to the metal, a N-Si $\pi*$ orbital is
populated with metal d-electron density. In the second form of the
π bond to the metal, the N-Si π bonding orbital is depopulated.
Either way, the formation of a π bond between the metal and the N
atom reduces the Si-N bond order. Sc^{III}, having no d electrons, is
unable to donate π electrons to N, and instead, the amide ligands in
the Sc^{III} compound are π donors to the metal.

(f) The orbital diagrams are shown in the answer for (e) above. For scandium, it is apparent that the amide ligand must be the π donor, as discussed above in (e). For the other metals, the authors conclude that it is not possible to say "whether the metal atoms are behaving as π donors."

(g) A likely reaction of these amide complexes with water would be hydrolysis of the M-N bonds to give bis(trimethylsilyl)amines and metal hydroxides or oxides.

A - Review

11-1A

The important minerals of the elements from Group IIA(2) are:

    beryl, $Be_3Al_2(SiO_3)_6$

    limestone, $CaCO_3$

    dolomite, $CaCO_3 \cdot MgCO_3$

    carnallite, $KCl \cdot MgCl_2 \cdot 6H_2O$

    strontianite, $SrSO_4$

    barytes, $BaSO_4$

11-3A

Beryllium has a small atomic radius and a high ionization enthalpy. The lattice energies of its salts and the hydration energies of its aqueous solutions are not high enough to favor the formation or stabilization of purely ionic compounds of beryllium. Consequently even compounds of beryllium with highly electronegative elements have large covalent character.

11-5A

Beryllium salts dissolve in water to give very acidic solutions due to hydrolysis as in equation 11-2.1. This is because the charge density at Be^{2+} is so high. Ca^{2+} solutions are not so extensively hydrolyzed.

11-7A

Three preparations of magnesium metal should be mentioned.

(a)   electrolysis of fused halide mixtures:

$$MgCl_2 \rightarrow Mg + Cl_2$$

(b)   reduction of the oxide with coke:

$$MgO + C \rightarrow Mg + CO$$

(c) calcination of dolomite:

$$MgCO_3 \cdot CaCO_3 \rightarrow MgO \cdot CaO + 2CO_3$$

followed by reduction as in equation 11-3.2.

11-9A
The solubilities of the chlorides and sulfates of the Group IIA(2) metals decrease from top to bottom in the group. The fluorides display an inverse solubility order because of the small size of the fluoride anion. The hydroxides of the Group IIA(2) metals increase in solubility from top to bottom in the group.

11-11A
Compounds of beryllium are toxic.

11-13A
Mg and Ca form complexes, mostly of oxygen donor ligands such as ethers, alcohols, cyclic polyethers such as crown ethers, and $EDTA^{4-}$. Water forms a six-coordinate complex of Mg^{2+}, the hexaaquo ion $[Mg(H_2O)_6]^{2+}$, which, unlike the tetraaquo complex of Be^{2+}, is not acidic.

B - Additional Exercises

11-1B
See Figure 11-2.

11-3B
The Mg^{2+} ion is considerably larger than that of beryllium, and the charge density at Be^{2+} is much greater. Only four ligands can be accommodated at Be^{2+}, whereas six ligands can fit around the larger magnesium ion.

11-5B
The metallic bond of the Group IIA(2) elements is stronger than that of the Group IA(1) elements because the former have a second valence

electron available for delocalization into the metallic network. The metal atoms are bound more tightly together in the Group IIA(2) elements, and the physical properties which depend on the extent of interatomic interaction reflect this stronger bond.

11-7B

Vapors of $BeMe_2$ are composed of linear molecules in which Be is sp-hybridized. The solid, on the other hand, forms chain polymers similar to that of $[BeCl_2]_n$, whose structure is shown in Figure 11-1. The bonding in the Be-Me-Be bridges is best described as a three-center, two-electron bond. We consider that the alkyl group is an anionic two-electron contributor to the three center bond. The orbital overlap picture and the MO energy level diagram are very much similar to those drawn in Chapter 3 and in Figure 12-5 for diborane, which also features a three-center, two-electron bond. The orbitals which the bridging groups contribute to the three center bond are different, but the overlap principle is the same. The bridging methyl group contributes an sp^3 hybrid orbital. The H atom in diborane contributes an atomic 1s orbital.

11-9B

11-11B

(a) $BeCl_2 + Mg \rightarrow Be + MgCl_2$

(b) $4Be + 1/2O_2 + N_2 \rightarrow BeO + Be_3N_2$

(c) $Be + 2OH^- + 2H_2O \rightarrow Be(OH)_4^{2-} + H_2$

(d) $Be(OH)_2 + 2OH^- \rightarrow Be(OH)_4^{2-}$

(e) $[Be(NH_3)_4]Cl_2 + 4H_2O + 4H^+ \rightarrow Be(H_2O)_4^{2+} + 2Cl^- + 4NH_4^+$

(f) $Be(H_2O)_4^{2+} + 4F^- \rightarrow BeF_4^{2-} + 4H_2O$

(g) $BeO + 2NH_4HF_2 \rightarrow BeF_4^{2-} + H_2O + 2NH_4^+$
   $Be(OH)_2 + 2NH_4HF_2 \rightarrow BeF_4^{2-} + 2H_2O + 2NH_4^+$

(h) $(NH_4)_2BeF_4 \rightarrow BeF_2 + 2HF + 2NH_3$

(i) $BeCl_2 + 4H_2O \rightarrow Be(H_2O)_4^{2+} + 2Cl^-$

(j) $4Be(OH)_2 + 6CH_3CO_2H \rightarrow OBe_4(O_2CCH_3)_6 + 7H_2O$

C - Questions from the Literature of Inorganic Chemistry

11-1C

(a) The three separate series of stability constants for formation of complexes with ions of the alkaline earth metals are listed in the introduction to this paper as cases (a), (b) and (c). The log K series are:

(a) Mg > Ca > Sr > Ba.

(b) Mg < Ca < Sr < Ba.

(c) Mg < Ca > Sr > Ba.

(b) Series (a) complexes with the alkaline earth metal ions are distinguished by a decrease in log K values with an increase in the radii of the cations. As the cations become larger (from top to bottom in the group), the complexes in aqueous solution become less stable (lower values of log K). This occurs mainly for complexes of the alkaline earth cations with small or highly charged ions.

Series (b) complexes with the alkaline earth metal ions are distinguished by an increase in log K values with an increase in the radii of the cations. As the cations become larger (from top to bottom in the group), the complexes in aqueous solution become more stable (larger values of log K). This occurs mainly for the complexes of the alkaline earth cations with large oxyanions such as sulfate.

Series (c) complexes with the alkaline earth metal ions are distinguished by an irregular trend in log K values, and this occurs primarily with carboxylate ligands, a good example being the one studied in this paper, the malate anion as the sodium salt.

(c) The complexes of the alkaline earth cations and various ligands fit into categories (a), (b) or (c) as follows:

iminodiacetate, series (a).

thiosulfate, series (b).

sulfate, series (b).

malate, series (c).

This conclusion can be reached by examining the log K values reported either in Table II, Table III or references 8 and 9.

(d) Entropy is a controlling factor for the series of complexes with iminodiacetate, for which the stability order of the complexes follows the same order as the $\Delta S°$ values, but is opposed to the order followed by the $\Delta H°$ values (Table III). The positive values of $\Delta S°$ reflect a large loss in order among the solvent molecules around the cations in aqueous solution - molecules which must be displaced and disrupted upon complexation of the aquated cations by the iminodiacetate ligand. Notice that a different conclusion must be reached for the trends in stability constants involving the malate ion (Table II). Here it is $\Delta H°$ that governs trends in log K. The author argues that, in contrast to the series involving iminodiacetate, the malate complexes offer a situation where metal-to-ligand bond strength is the principal issue controlling the stability of the complexes.

CHAPTER 12

A - Review

12-1A

This is the icosahedron, shown in structure 8-II.

12-3A

Boric acid, $B(OH)_3$, ionizes in water as in equation 12-3.2. Based on the value of pK_a, we can judge that it is a weak acid, comparable to phenol or HCN.

12-5A

BF_3 may be prepared by reacting the oxide in sulfuric acid solution with a fluoride source, either NH_4BF_4 or CaF_2:

$$B_2O_3 + 6BF_4^- + 6H^+ \rightarrow 8BF_3 + 3H_2O$$

12-7A

The structure of diborane is given in Figures 12-4 and 12-5, and the bonding is described thoroughly in Chapter 3, especially Figures 3-33, 3-34, and 3-35.

12-9A

A number of hybridoborates can be prepared, some typical reactions being 12-5.24 through 12-5.28, all of which feature the use of hydride transfer agents.

B - Additional Exercises

12-1B

Of the hydrides of boron that are listed in Table 12-1 and in Figure 12-4, none is properly considered to be either an arachno or a nido substructure of an icosahedron. However, $B_{12}H_{12}^{2-}$ does have the geometry of an icosahedron.

12-3B

Some reactions of diborane are diagramed below. The reactions are
not balanced. Often other products (for instance, H_2) are formed
along with the principal products that are listed.

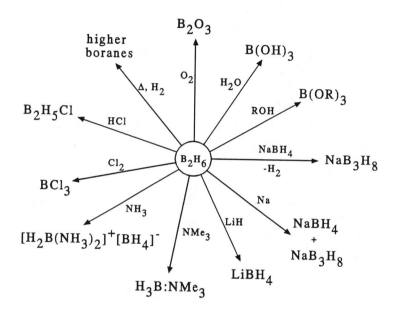

12-5B

The structure of the anion $[B_3H_8]^-$ is given below:

$$\left[\begin{array}{c} H \diagdown \quad \diagup H \\ H{-}B{-}H \\ \\ H-B \qquad\qquad B-H \\ | \qquad\qquad | \\ H \qquad\qquad H \end{array} \right]^{-} \longleftrightarrow \left[\begin{array}{c} H \diagdown \quad \diagup H \\ H{-}B{-}H \\ \\ H-B \qquad\qquad B-H \\ | \qquad\qquad | \\ H \qquad\qquad H \end{array} \right]^{-}$$

12-7B

These should be considered to be derivatives of boric acid, each containing trigonal (sp^2-hybridized) boron. Thus, we have
(a) orthoborates, $B(OR)_3$; (b) acylborates, $B[O(C=O)R]_3$; and
(c) peroxoborates, $B(OOR)_3$.

(a) $B(OR)_3$. The oxygen atoms are of structural type AB_2E_2. The hybridization at oxygen is sp^3, and the BOR geometry is bent:

$$B \longrightarrow \underset{..}{\overset{}{O}}{:} \diagup^{R}$$

(b) $B[O(C=O)R]_3$. The carbonyl carbon atom is of the type AB_3 and it is sp^2-hybridized. The oxygen atom bound to boron is of the type AB_2E_2, and it is sp^3 hybridized:

$$\begin{array}{c} :\overset{..}{O} \\ \| \\ C - R \\ B \longrightarrow \underset{..}{O}{:} \end{array}$$

(c) $B(OOR)_3$. Each oxygen is of the type AB_2E_2 and is sp^3-hybridized:

$$\begin{array}{c} :\overset{..}{\underset{..}{O}}{-}R \\ B \longrightarrow \underset{..}{O}{:} \end{array}$$

12-9B

(a) $BF_3 + OEt_2 \rightarrow F_3B:OEt_2$

(b) See equations 12-4.1 and 12-4.2.

(c) $BCl_3 + 3ROH \rightarrow B(OR)_3 + 3HCl$

(d) $B_2H_6 + HCl \rightarrow B_2H_5Cl + H_2$

(e) $B_{10}H_{14} + NR_3 \rightarrow [R_3NH][B_{10}H_{13}]$

(f) $B_{10}H_{14} + I_2 \rightarrow 2,4-I_2B_{10}H_{12} + H_2$

(g) $2LiH + B_2H_6 \rightarrow 2LiBH_4$

(h) $NH_4Cl + LiBH_4 \rightarrow H_3B:NH_3 + H_2 + LiCl$

(i) $Me_2N=BCl_2 + 2PhMgBr \rightarrow Me_2N=BPh_2 + 2"MgBrCl"$

(j) $B_3N_3H_6 + 9H_2O \rightarrow 3NH_3 + 3B(OH)_3 + 3H_2$

(k) $B_3N_3H_6 + 3HBr \rightarrow [-H_2N-BHBr-]_3$

(l) $(Cl-B)_3(NH)_3 + 3EtMgBr \rightarrow (Et-B)_3(NH)_3 + 3"MgBrCl"$

12-11B

Addition of HX to borazine takes place readily because of the polarity in the B-N bonds of borazine. There is no such bond polarity in benzene.

12-13B

We must account for a total of $(3 \times 10 + 14) = 44$ valence electrons. They are distributed as follows:

Structural Feature	Number of Occurrences	Number of Electrons
terminal, 2c-2e B-H groups	10	20
bridging, 3c-2e B-H-B groups	4	8
open, 3c-2e B-B-B bridges	2	4
closed, 3c-2e B-B-B bonds	4	8
2c-2e BB bonds	2	4
		total = 44

12-15B

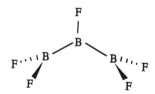

12-17B

(a) $B_2H_6 + 2NH_3 \rightarrow [H_2B(NH_3)_2]^+[BH_4]^-$

(b) $B_2H_6 + HCl \rightarrow B_2H_5Cl + H_2$

(c) $B_2O_3 + 2Fe \rightarrow 2B + Fe_2O_3$

(d) $B(OH)_3 + 3CH_3C(=O)Cl \rightarrow B(OC(=O)CH_3)_3 + 3HCl$

(e) $2BF_3 + 6NaH \rightarrow B_2H_6 + 6NaF$

(f) $B_{10}H_{14} + I_2 \rightarrow 2,4-I_2B_{10}H_{12} + H_2$

(g) $B_5H_9 + 15H_2O \rightarrow 5B(OH)_3 + 12H_2$

(h) $B_2H_6 + 3O_2 \rightarrow B_2O_3 + 3H_2O$

(i) $BCl_3 + 3H_2O \rightarrow B(OH)_3 + 3HCl$

(j) $B(OH)_3 + 2EtOH \rightarrow B(OEt)_3 + 3H_2O$

(k) no reaction

(l) $2B_2H_6 + 2Na \rightarrow NaBH_4 + NaB_3H_8$

(m) $B_2H_6 + 6H_2O \rightarrow 2B(OH)_3 + 6H_2$

(n) $BCl_3 + 3EtOH \rightarrow B(OEt)_3 + 3HCl$

(o) $2B(OH)_3 \rightarrow B_2O_3 + 3H_2O$

(p) $B(OH)_3 + 2NH_4HF_2 \rightarrow NH_4BF_4 + 2H_2O + NH_4OH$

12-19B

(a) $2BF_3 + 6NaH \rightarrow B_2H_6 + 6NaF$

   or

   $4BF_3 + 3NaBH_4 \rightarrow 2B_2H_6 + 3NaBF_4$

(b) $2NH_4Cl + 3BCl_3 \rightarrow [ClB-NH]_3 + 9HCl$

(c) $B(OH)_3 + 3MeOH \rightarrow B(OMe)_3 + 3H_2O$

(d) $3B_2Cl_4 + 4SbF_3 \rightarrow 3B_2F_4 + 4SbCl_3$

(e) $2BCl_3 + 2Hg \rightarrow B_2Cl_4 + Hg_2Cl_2$

   (only under electric discharge)

(f) BCl condenses at -196 °C in the presence of BCl_3 to give B_4Cl_4:

$$4BCl \rightarrow B_4Cl_4$$

Small amounts of B_4Cl_4 can also be made directly from BCl_3 by radio frequency dicsharge in the presence of mercury.

(g) B_2Cl_4 decomposes thermally to give a mixture of halides, which must be separated:

$$B_2Cl_4 \rightarrow B_8Cl_8 + B_9Cl_9 + B_{10}Cl_{10} + B_{11}Cl_{11}$$

(h) The first step is <u>not</u> a recommended route to $B(OH)_3$:

$$B_2H_6 + 6H_2O \rightarrow 2B(OH)_3 + 6H_2$$
$$B(OH)_3 + 3EtOH \rightarrow B(OEt)_3 + 3H_2O$$

(i) $BF_3 + 1/2Al_2Cl_6 \rightarrow AlF_3 + BCl_3$

(j) $2B_2H_6 + 2Na \rightarrow NaBH_4 + NaB_3H_8$

(k) $3NH_4Cl + 3BCl_3 \rightarrow [ClB-NH]_3 + 9HCl$

$$[ClB-NH]_3 + 3NaBH_4 \rightarrow 3NaCl + B_3N_3H_6 + 3/2B_2H_6$$

C - Problems from the Literature of Inorganic Chemistry

12-1C

(a) With a general nucleophile L we have symmetrical cleavage:

$$B_4H_{10} + 2L \rightarrow H_3B:L + LB_3H_7$$

or unsymmetrical cleavage:

$$B_4H_{10} + 2L \rightarrow [H_2BL_2]^+[B_3H_8]^-$$

(b) With sodium borohydride we have symmetrical cleavage:

$$B_4H_{10} + NaBH_4 \rightarrow NaB_3H_8 + B_2H_6$$

and unsymmetrical cleavage:

$$B_4H_{10} + NaBH_4 \rightarrow NaB_3H_8 + B_2H_6$$

In the absence of the sort of labeling data that are presented in this paper, the two possibilities in (b) are clearly indistinguishable.

(c) In symmetrical cleavage, $[BD_4]^-$ should act as a $[D]^-$ donor and as a source of "BD_3":

$$B_4H_{10} + [BD_4]^- \rightarrow [B_3H_7D]^- + B_2H_3D_3$$

In unsymmetrical cleavage, $[BD_4]^-$ should remove a $[BH_2]^+$ unit from B_4H_{10} to give diborane:

$$B_4H_{10} + [BD_4]^- \rightarrow [B_3H_8]^- + B_2H_2D_4$$

The %D label can be calculated based on the ratio of D to (D + H) found in each possible product. The percent D found in the product should allow for a distinction between the two possible reaction pathways.

(d) Exchange of H and D among the reactants or products would make the results meaningless, because the experimental %D found in any of the products would not then depend on the reaction pathway.

(e) Ammonia gives unsymmetrical cleavage of tetraborane(10):

$$B_4H_{10} + 2NH_3 \rightarrow [H_2B(NH_3)_2][B_3H_8]$$

12-3C

(a) (i) Monomethylation takes place at the 5-boron atom, giving $5\text{-}MeC_2B_5H_6$.

   (ii) Dimethylation takes place at the 5- and 6-boron atoms, giving $5,6\text{-}Me_2C_2B_5H_5$.

  (iii) Trimethylation gave the $1,5,6\text{-}Me_3C_2B_5H_4$ isomer.

(b) The positional preferences for Friedel-Crafts methylation are 5,6 > 1,7 > 3, indicating that electrons are most available for electrophilic substitution at particular boron atoms in the order 5,6 > 1,7 > 3.

(c) The belt carbon-atoms in $2,4\text{-}C_2B_5H_7$ are ascribed electropositive character because they are not methylated in these Friedel-Crafts reactions. We can then divide the boron atoms into three groups. Belt-boron B(3) is unique, being adjacent to both belt-carbon atoms.

Belt-boron atoms B(5) and B(6) are each adjacent to only one carbon atom. Apical-boron atoms B(1) and B(7) have long bonds to each belt-carbon atom. The availability of electron density for Friedel-Crafts alkylation at a particular boron atom then depends on the position of the boron atom relative to the electropositive carbon atoms. B(5) and B(6) are relatively electron-rich because they are each adjacent to only one electron-withdrawing carbon atom. B(1) and B(7) each have long bonds to both electron-withdrawing carbon atoms. B(3) is least electron-rich, having short bonds to both belt-carbon atoms.

12-5C

(a) Reduction of B_5H_9 was performed using an alkali metal napthalenide reagent:

$$B_5H_9 + 2M^+[C_{10}H_8]^- \rightarrow M_2B_5H_9 + C_{10}H_8$$

Protonation of $M_2B_5H_9$ with anhydrous HCl gave B_5H_{11}:

$$M_2B_5H_9 + 2HCl \rightarrow 2MCl + B_5H_{11}$$

(b) For B_5H_9, $F = 3(5) + 9 - 2(5) = 14$. Since $14 = 2n + 4$, we arrive at a nido structure.

For $B_5H_9{}^{2-}$, $F = 3(5) + 9 + 2 - 2(5) = 16$. Since $16 = 2n + 6$, we arrive at an arachno structure.

For B_5H_{11}, $F = 3(5) + 11 - 2(5) = 16$. Since $16 = 2n + 6$, we arrive at an arachno structure.

(c) $B_5H_9{}^{2-}$ and $B_5H_8{}^-$ both show only two ^{11}B NMR resonances, even at low temperature, suggesting that they are both fluxional. Otherwise, the stabilities, insolubilities, and IR spectra of salts of these two anions are different. Also, protonation of $B_5H_8{}^-$ gives B_5H_9, whereas protonation of $B_5H_9{}^{2-}$ gives B_5H_{11}.

A - Review

13-1A

Bauxite is the readily available ore of aluminum. It is variously composed as the hydrous oxide, $Al_2O_3 \cdot nH_2O$, where n = 1 - 3. The ore is purified by dissolution in aqueous NaOH, followed by reprecipitation of the hydroxide, $Al(OH)_3$. The anhydrous hydroxide, which is obtained after thorough dehydration of the hydroxide, is dissolved in molten cryolite (or in synthetic Na_3AlF_6), and the melt is electrolyzed to obtain aluminum metal.

13-3A

(a) Corundum is anhydrous α-Al_2O_3. Its structure, an hcp array of oxide ions with aluminum ions in two-thirds of the octahedral holes, is discussed in Section 4-7.

(b) Spinel is the mineral $MgAl_2O_4$, a mixed-metal oxide whose structure is discussed in Section 4-8. The oxide ions form a ccp array, and magnesium ions occupy a specific set of tetrahedral holes, while aluminum ions occupy a set of octahedral holes.

13-5A

The term alum refers to a great variety of salts with the general formula $M^IM^{III}(SO_4)_2 \cdot 12H_2O$. The prototype, aluminum alum, contains trivalent Al^{3+} as the hexaaquo ion ($[Al(H_2O)_6]^{3+}$), as well as the hexaaquo ion ($[M(H_2O)_6]^+$) of practically any monovalent cation other than Li^+. The sulfate anions and waters of hydration finish the structural components of alums. Metals besides Al^{3+} which are found in other alums are listed in Section 13-5.

13-7A

Fusion of boric acid, $B(OH)_3$, gives the anhydrous glass B_2O_3. Fusion or dehydration of the hydrous oxides or hydroxides of aluminum gives either α-Al_2O_3 (obtained above 1000 °C) or γ-Al_2O_3 (obtained at lower temperatures, 450 °C). As shown in equation 12-3.2, $B(OH)_3$ is acidic. $Al(OH)_3$ is amphoteric, as in equations 13-5.3 and 13-5.4.

13-9A

The amphoterism of the hydroxides of Al and Ga is represented by equations 13-5.3 through 13-5.6. Also, both the oxides and the hydroxides of Al and Ga dissolve both in aqueous acids and in aqueous bases. The more metallic members of the elements of Group IIIB(13), namely indium and thallium, give only basic oxides and hydroxides.

B - Additional Exercises

13-1B

Thallium(III) is more of an oxidizing ion than the other ions of the Group IIIB(13) metals. Also, iodide is the best reducing agent of the halide ions. The oxidizing strength of Tl^{III} is also shown by the ready decompositions, mentioned in Section 13-4 of $TlCl_3$ (at 40 °C) and $TlBr_3$ (below 40 °C):

$$TlCl_3 \rightarrow Cl_2 + TlCl$$
$$2TlBr_3 \rightarrow Br_2 + Tl[TlBr_4]$$

13-3B

The tetramer is shown in Structure 13-V. Note that the central aluminum atom is essentially six-coordinate, whereas the other three aluminum atoms are four-coordinate. Note also that there are two types of alkoxide ligands, bridging ones and terminal ones, which are distinguishable using nmr techniques.

13-5B

The reduction of Tl^{3+} to Tl^+ is sensitive to the presence of coordinating anions because the Tl^{III} state is stabilized by formation of four-coordinate complexes such as $[TlI_4]^-$. Aqueous Tl^{III} is hydrolyzed to give $[Tl(OH)]^{2+}$, and at high pH, the insoluble oxide:

$$2TlI_3 + 6OH^- \rightarrow Tl_2O_3 + 6I^- + 3H_2O$$

13-7B

(a) $AlCl_3 + PF_3 \rightarrow Cl_3Al:PF_3$

(b) $Li + 1/2\ H_2 \rightarrow LiH$

$Al + 3/2Cl_2 \rightarrow AlCl_3$

$4LiH + AlCl_3 \rightarrow LiAlH_4 + 3LiCl$

(c) $LiGaH_4 \rightarrow LiH + Ga + 2H_2$

(d) $TlCl_3 \rightarrow TlCl + Cl_2$

(e) $[Al(H_2O)_6]^{3+} = [Al(H_2O)_5OH]^{2+} + H^+$

(f) $GaCl_3 + 3H_2O \rightarrow Ga(OH)_3 + 2HCl$

(Also, note equilibria 13-5.5 and 13-5.6 in the text.)

(g) $Al + 3EtOH \rightarrow Al(OEt)_3 + 3/2H_2$

(h) $Tl_2O_3 \rightarrow Tl_2O + O_2$

(i) $Al_2Cl_6(g) + 2NMe_3 \rightarrow 2Cl_3Al:NMe_3$

(j) These are equations 13-5.3 and 13-5.4 in the text.

(k) $AlCl_3(s) + 2Al(s) \rightarrow 3AlCl(g)$

(l) $Me_3N:GaH_3 + NMe_3 \rightarrow H_3Ga(NMe_3)_2$

(This diadduct is stable only below -60 °C. See equations 13-7.6 through 13-7.9 in the text.)

C - Questions from the Literature of Inorganic Chemistry

13-1C

(a) The adducts $TlX_3\cdot2L$ and $InCl_3\cdot3L$ were prepared by simple addition reactions in methylcyanide solvents, for most of the ligands reported here:

$$TlX_3 + 2L \rightarrow TlX_3\cdot2L$$
$$InCl_3 + 3L \rightarrow InCl_3\cdot3L$$

The exception to the above stoichiometries is $TlCl_3\cdot3(\gamma\text{-picoline})$.

The trihalides of Tl were prepared by halogen-oxidation of the appropriate Tl^I salts in methyl cyanide solvent:

$$TlCl + Cl_2 \rightarrow TlCl_3$$
$$TlBr + Br_2 \rightarrow TlBr_3$$

(b) The infrared spectroscopic results aid in the structural assignments.

(i) The structure of $[TlCl_4]^-$ is most likely tetrahedral, although it seems to depend on the nature of the counter cation in the solid state, and on the solvent, when dissolved.

(ii) $TlCl_3 \cdot 2py$ is probably a monomeric, five-coordinate complex in non-coordinating solvents, but there is a general tendency for these systems to add another ligand (or solvent), or to aggregate in the solid state, so as to become six-coordinate.   For more information see:

1. F. A. Cotton, B. F. G. Johnson, and R. M. Wing, *Inorg. Chem.* **1965**, *4*, 502-507.

2. R. A. Walton, *J. Inorg. Nucl. Chem.* **1970**, *32*, 2875-2884.

3. J. R. Hudman, M. Patel, and W. R. McWhinnie, *Inorg. Chim. Acta* **1970**, *4*, 161-165.

4. R. A. Walton, *Coord. Chem. Rev.*, **1971**, *6*, 1-25.

(iii) $InCl_3 \cdot 3py$ is likely monomeric, six-coordinate, and pseudo-octahedral.

13-3C

(a)

(i) BF_3 forms no adduct reported in this paper, although BH_3 is known to form the adduct $H_3B \leftarrow PF_3$.

(ii) Al_2Cl_6 gives $Cl_3Al \leftarrow L$ adducts with L = PF_3, but not with L = CO or PCl_3, the latter presumably because of chlorine-chlorine repulsions which prevent close approach between P and Al.

(iii) $Al_2(CH_3)_6$ gives $(CH_3)_3Al \leftarrow L$ adducts with $L = PF_3$ and NH_3, but not with $L = CO$ or PCl_3. Evidently a methyl group is about as large as a chloride.

(b) The authors conclude that PF_3 is a better σ donor than CO, because where there is room to do so, PF_3 forms an adduct (with $AlCl_3$ and with $Al(CH_3)_3$) whereas CO does not.

(c) BF_3 is planar; boron is sp^2-hybridized (class AB_3 according to the occupancy scheme of Chapter 3), and there is extensive intramolecular $p\pi$-$p\pi$ bonding between B and F. The "distortion energy" that is required in order to permit adduct ($F_3B \leftarrow L$) formation is that energy associated with going to tetrahedral, sp^3-hybridized, (non-π-bonded) boron in the adduct.

"$AlCl_3$", which is really Al_2Cl_6, already has an sp^3-hybridized aluminum atom, and less distortion energy is required upon adduct formation. Also, Al, being larger than B, is more able sterically to accommodate coordination number four.

CHAPTER 14

A - Review

14-1A
Carbon is normally considered to be promoted to the valence state $[2s^1 2p_x^1 2p_y^1 2p_z^1]$, from which it is possible to construct the familiar four-valence hybridizations: sp^3 (AB_4 systems such as CCl_4), sp^2 (AB_3 systems such as CO_3^{2-} and Cl_2CO), and sp (AB_2 systems such as CN^-, CO_2 and CS_2).

14-3A
Catenation is self-binding: formation of chains or rings between like atoms, using single or multiple bonds. Carbon does it most widely, the chain and cyclic hydrocarbons providing familiar examples. Catenation is prevalent in sulfur chemistry, as discussed in Section 19-1. For example, elemental sulfur occurs as catenated S_8 rings. The great strength of the carbon-carbon bond is responsible for the tendency toward catenation for carbon. The tendency toward catenation is much reduced in other elements of Group IVB(14). For silicon, although the silicon-silicon single bond is strong (226 kJ mol^{-1}), the Si-O bond is especially strong (368 kJ mol^{-1}), such that catenated Si-Si compounds are unstable in the presence of oxygen, giving Si-O linkages.

14-5A
Whether finely divided (soot and lampblack) or greatly aggregated and apparently amorphous (charcoals or block graphite), most carbon other than diamond is composed of microcrystalline particles, each having the graphite structure or the fullerene structures discussed in Section 8-5. Graphite is less dense than diamond, and its layered structure allows for ready cleavage between the planes.

14-7A
Carbon dioxide is available from complete combustion of hydrocarbons:

$$C_6H_{14} + 9.5O_2 \rightarrow 6CO_2 + 7H_2O$$

from calcination of limestone, or heating of carbonates in general:

$$CaCO_3 \rightarrow CO_2 + CaO$$

and from the treatment of aqueous solutions of simple carbonate salts, or of solid carbonates, with acids:

$$CO_3^{2-} + 2H^+ = H_2CO_3 \rightarrow CO_2 + H_2O$$
$$SrCO_3 + 2HCl \rightarrow CO_2 + H_2O + SrCl_2$$

14-9A
We can view this as a neutralization represented by the following equilibrium:

$$CO_2 + CO_3^{2-} + H_2O = 2HCO_3^-$$

14-11A
Both $(CN)_2$ and Cl_2 are oxidizing agents, being reduced, to CN^- and Cl^-, respectively. Both Cl_2 and $(CN)_2$ disproportionate in basic solution:

$$(CN)_2 + 2OH^- \rightarrow CN^- + OCN^- + H_2O$$
$$Cl_2 + 2OH^- \rightarrow Cl^- + OCl^- + H_2O$$

Cyanogen and chlorine undergo oxidative addition reactions with lower-valent transition metal complexes, as described in Chapter 30 and in equation 14-5.5. Other similarities with halogens in general are listed in Section 14-5.

14-13A
The industrial synthesis of HCN is accomplished by reaction 14-5.7 or 14-5.8. HCN is a toxic, low boiling liquid that is a strong acid, although its acidity is leveled in water solution. The high dielectric constant of HCN makes it a good solvent for ionic substances. Other properties of HCN are listed in Section 14-5.

14-15A

(a) This can be accomplished by liberation of carbon dioxide, reduction to carbon monoxide, and complexation with zerovalent nickel:

$$BaCO_3 + 2H^+ \rightarrow Ba^{2+} + CO_2$$
$$CO_2 + Zn \rightarrow ZnO + CO$$
$$Ni + 4CO \rightarrow Ni(CO)_4$$

(b) The carbonate can be reduced to a carbide, and the carbide reacted with water:

$$BaC_2 + H_2O \rightarrow C_2H_2 + BaO$$

(c) The carbonate can be treated with acid to give CO_2, which is reduced by hydrogen:

$$CO_2 + 4H_2 \rightarrow CH_4 + 2H_2O$$

(d) Methane from (c) above is reacted as in equation 14-6.1 to give labeled CS_2:

$$CH_4 + 4S \rightarrow CS_2 + 2H_2S$$

(e) The carbonate is treated with acid to give CO_2, and the CO_2 can be reduced to CO. Various catalytic processes can lead eventually to methanol synthesis:

$$CO + 2H_2 \rightarrow CH_3OH$$

B - Additional Exercises

14-1B

Substance	Bond Distance	Bond Order
ethylene	1.33	2.0
benzene	1.39	1.5
graphite	1.42	1.3
diamond	1.54	1.0

Considering a single layer of the graphite structure, we have everywhere, sp^2-hybridized carbon atoms of the AB_3 classification. The geometry at each carbon is trigonal, and each carbon participates in two single and one double bond. The effect of resonance, or of delocalization of π electron density, is that the double bond is distributed evenly to each of the three carbon atoms bonded to any given carbon atom. The bond order is therefore 1/3 greater than 1.0, *i.e.* 1.33.

14-3B

Dimers:

Trimers:

Tetramer:

Pentamer:

Polymer:

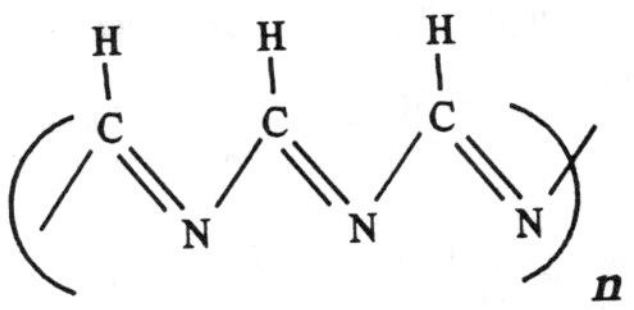

14-5B

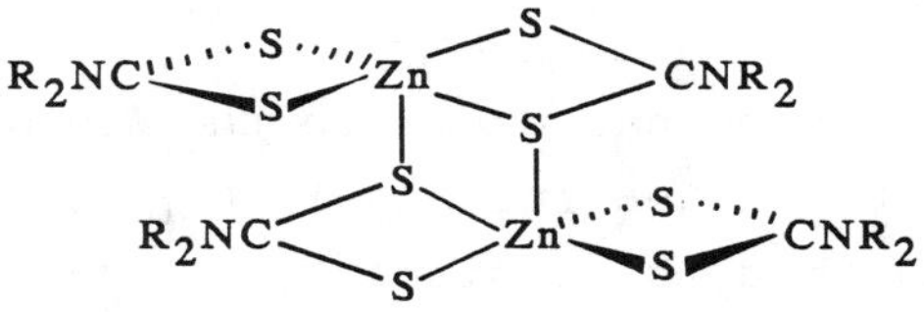

14-7B

For reaction 14-5.1, HCN is the reducing agent and NO_2 is the oxidizing agent. The Lewis diagrams are:

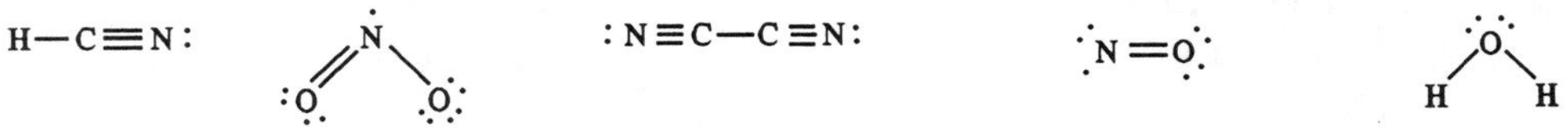

For reaction 14-5.2, NO is the reducing agent, and O_2 is the oxidizing agent. The Lewis diagrams are:

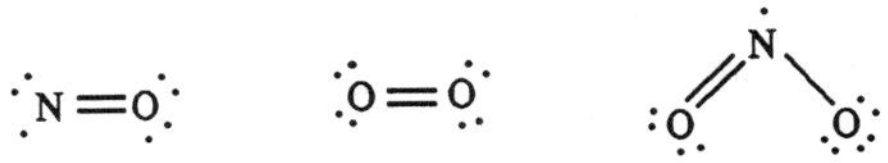

The Lewis diagrams ought to help to establish, through electron book-keeping, the redox nature of these and other reactions.

C - Questions from the Literature of Inorganic Chemistry

14-1C

(a) The Ir atom is attached to one C-C bond in particular, a 6-6 ring fusion site. This results in C_1 and C_2 being pulled away from the

C_{60} sphereoid, toward the Ir atom. The distance between C_1 and C_2 is 1.53 Å, which is one of the longest C-C distances in the C_{60} unit.

(b) When the title compound is dissolved in CH_2Cl_2, the IR and ^{31}P NMR spectra are found to be simply those of the separate molecules $Ir(CO)Cl(PPh_3)_2$ and C_{60}.

(c) We conclude that a 6-6 ring fusion has added to the Ir atom, not a 6-5 fusion. This suggests that the 6-6 fusions in C_{60} are more electron deficient, a conclusion that is consistent with theory.

(d) The benzene molecules in the crystal do not interact with the Ir-C_{60} complex. They merely fill space between the molecules of the crystal. In other words, the title compound is a hexa-solvated crystalline substance. The authors point out that this may be an important factor in their having been able to isolate this very weak adduct as a crystalline solid.

A - Review

15-1A

Carbon dioxide is a nonpolar, molecular substance, in which intermolecular forces of attraction (van der Waals forces) are weak. SiO_2 is a network covalent substance, composed of continuously linked tetrahedra.

15-3A

Recall the material of Section 8-11. Although a number of SiX_2 substances are known, they are generally unstable. Germanium dihalides are more stable than the silicon analogues, although $GeCl_2$ does readily add Cl_2 to give the tetravalent $GeCl_4$, as shown in equation 8-11.3. The relative stabilities of the divalent Group IVB(14) elements follow from their relative reactivities towards Cl_2, as shown by equations 8-11.2 through 8-11.4. Most likely it is the decrease in M-Cl bond energies from top to bottom in Group IVB(14) that makes formation of the tetravalent MCl_4 compounds unfavorable for those elements towards the bottom of the group.

15-5A

$$SiCl_4 + LiAlH_4 \rightarrow SiH_4 + LiCl + AlCl_3$$
$$SiH_4 + (n + 2)H_2O \rightarrow SiO_2 \cdot nH_2O + 2H_2$$

15-7A

The formation of the stable hexafluoro dianion according to equation 15-6.1 prevents complete hydrolysis of SiF_4. Note that other hexaflourides of Group IVB(14) elements are not stable in basic solution ($[GeF_6]^{2-}$ and $[SnF_6]^{2-}$) or even in water ($[PbF_6]^{2-}$). Hydrolysis leads to hydrous oxides or hydroxy-halides, rather than to simple hydroxides.

15-9A

Silicon shows less tendency towards catenation than does carbon because the Si-Si single bond energy (226 kJ mol^{-1}) is lower than that for carbon (356 kJ mol^{-1}) and because there is less ability on the part of silicon to form Si=Si or other types of multiple bonds of the $p\pi$-$p\pi$ variety. Nevertheless, a number of silicon compounds containing single and multiple bonds are now being reported. See the review articles listed under "Supplementary Reading."

15-11A
Red lead is the mixed oxide of lead, Pb_3O_4, containing chains of $Pb^{IV}O_6$ octahedra which share two oxygen atoms along a common edge. These chains are linked by pyramidal Pb^{II} ions which alternately bind to two oxygen atoms of one chain and one oxygen of another chain. Red lead is made by heating PbO (litharge, a red form with a tetragonal structure) with PbO_2 at 250 °C. Red lead may also be prepared by heating PbO in air at 450 °C. Numerous other forms of PbO and PbO_2 exist, as well as oxides of other stoichiometries.

B - Additional Exercises

15-1B
The nitrogen atom in H_3CNCS is of the AB_2E classification, and it is sp^2-hybridized:

$$\underset{H_3C}{\overset{\displaystyle \,}{\diagup}}\ddot{N}=C=\ddot{S}\!:$$

The nitrogen atom of H_3SiNCS is of the AB_2 classification, and it is sp-hybridized:

$$H_3Si=N=C=\ddot{S}\!:$$

This is made possible by $p\pi \rightarrow d\pi$ bonding in which $p\pi$ electron density on nitrogen is donated into an empty d orbital on Si. The molecule would not be linear at nitrogen if this π bonding were absent.

15-3B
From the various compounds mentioned in the chapter, especially Table 15-1, we can construct the following list:

Substances	AB_xE_y Classification of the Group IVB(14) Atom	Predicted Geometry at the Group IVB(14) Atom
$N(SiH_3)_3$	AB_4	Tetrahedral
$[Pt(SnCl_3)_5]^{3-}$	AB_4	Tetrahedral
MX_4, SiH_4, $HSiCl_3$	AB_4	Tetrahedral
$Si(OC_2H_5)_4$	AB_4	Tetrahedral
$SnCl_3^-$, $GeCl_3^-$	AB_3E	Pyramidal
$SnCl_2{\cdot}OC(CH_3)_2$	AB_3E	Pyramidal
$[SiF_6]^{2-}$	AB_6	Octahedral
$Sn(NO_3)_4$	AB_8	Dodecahedral

15-5B

The ion $[SnCl_3]^-$ contains a tin atom of the AB_3E classification. This sp^3-hybridized Sn atom has a lone pair of electrons that can be donated to metals.

15-7B

The $p\pi$-$p\pi$ bond between nitrogen and boron in the bis(*tert*-butyl) derivative should be much stronger than in the similar bis(trimethylsilyl) derivative because the latter offers the chance for $p\pi{\to}d\pi$ donation of electron density from nitrogen to silicon:

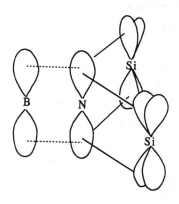

The competition between silicon and boron for π electron density on nitrogen weakens the π bond from nitrogen to boron in the silicon derivative.

15-9B

The AB_xE_y classification and the geometries at Sn are presented in problem 15-3B above. The Lewis diagrams are:

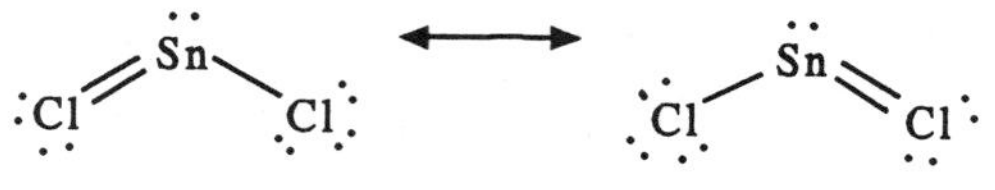

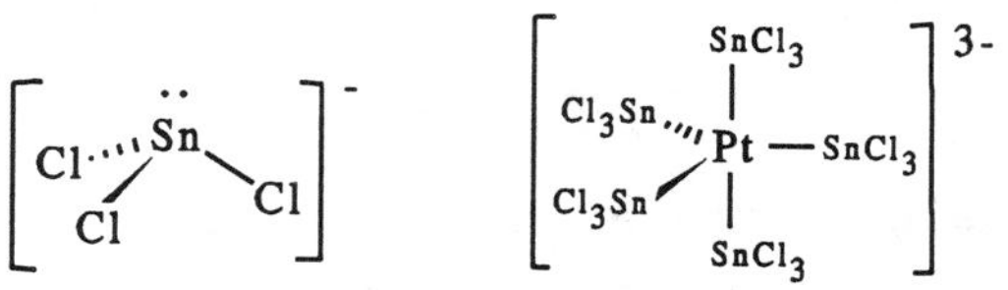

15-11B

The Cl-Sn-Cl angle in $SnCl_3^-$ (as the Cs^+ salt) is 87-92°. There is only a slight interaction between the Sn atom of one $SnCl_3$ pyramid and the Cl atoms of another in the solid state. The stable ion in aqueous solutions of SnF_2 and F^- is the pyramidal SnF_3^- ion. We expect the F-Sn-F angle to be less than the Cl-Sn-Cl angle of $SnCl_3^-$ because of the greater electronegativity of fluorine, which draws the electron density of the Sn-F bonds out towards fluorine, allowing some collapse of the F-Sn-F angle so as to alleviate repulsion between the Sn-F bonding pairs and the Sn lone electron pair. In the solid, di-tin and tri-tin ions such as $[Sn_2F_5]^-$ and $[Sn_3F_{10}]^{4-}$ are formed, so that a direct comparison of structural data is not appropriate. The solid structure of $Cs[GeCl_3]$ has sharing of chloride groups between different Ge atoms, to give roughly six-coordination for each germanium atom. The angles found here cannot, therefore, properly be compared with those of $Cs[SnCl_3]$, which does not have this aggregation. We can predict, though, that the $GeCl_3^-$ ion in solution should have Cl-Ge-Cl angles that are larger than those of $SnCl_3^-$, because Ge is a smaller atom than Sn; the Cl-Ge-Cl angles must become larger to alleviate repulsions among the three "close in" Cl-Ge bonding pairs of electrons. Remember too that the Ge lone electron pair is "smaller" than that of Sn. (Some of this reasoning is troublesome because there may be substantial differences in hybridization at the central atoms.)

15-13B

This is shown by the comparison provided by equations 8-11.2 through 8-11.5. The stability order is: $PbCl_2 > SnCl_2 > GeCl_2$.

15-15B

The Lewis diagram is:

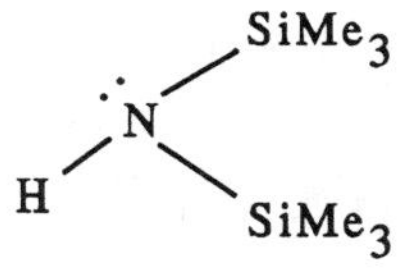

Each Si atom is tetrahedrally coordinated (AB_4). The nitrogen atom, however, is of the type AB_3E. This would normally give a pyramidal geometry at nitrogen, but the lone pair on the nitrogen atom is delocalized into the d orbitals of the silicon atoms, making the Si_2NH unit planar:

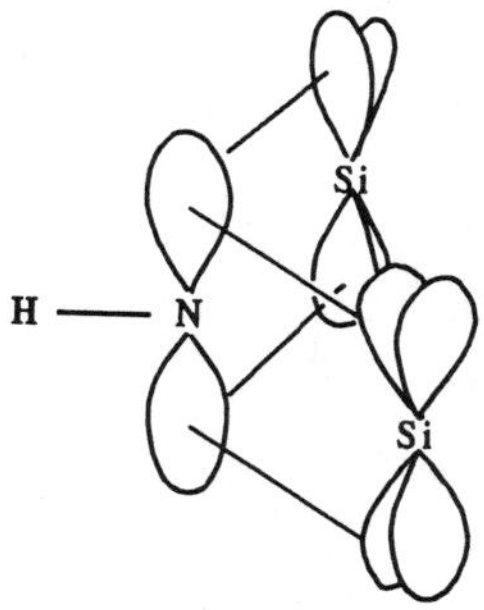

15-17B

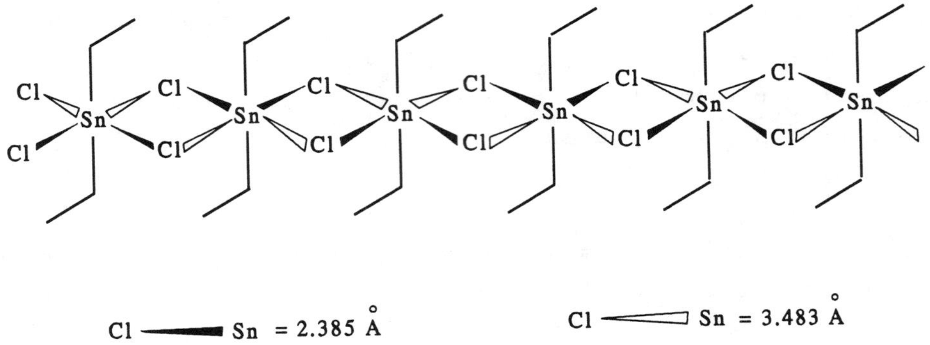

$$Cl \longrightarrow Sn = 2.385 \text{ Å} \qquad Cl \Longrightarrow Sn = 3.483 \text{ Å}$$

15-19B

The Lux-Flood base (oxide ion donor) in this reaction is Na_2O, which should be basic since it is the oxide of a metal. Conversely, SiO_2 is a Lux-Flood acid, namely an oxide ion acceptor.

15-21B

(a) $GeO_2 + C \rightarrow Ge + CO_2$

$GeO_2 + 2H_2 \rightarrow Ge + 2H_2O$

(b) $Sn + 2Cl_2 \rightarrow SnCl_4$

(c) $SnCl_2 + py \rightarrow Cl_2Sn:py$

(d) $GeCl_4 + xsH_2O \rightarrow Ge(OH)_4 \cdot nH_2O + 4HCl$

(e) See equations 15-7.4 and 15-7.5

(f) See equation 15-6.1

15-23B

Oxidation of Si gives SiO_2, which can then be treated with aqueous HF to give SiF_6^{2-} solutions.

15-25B

$$Li + (SiMe_3)_2NH \rightarrow Li^+(SiMe_3)_2N^- + 1/2H_2$$

C - Questions from the Literature of Inorganic Chemistry

15-1C

(a) We expect <u>mono</u>aminoboranes to have the highest B-N rotational barriers (17-24 kcal/mol), because only one nitrogen atom donates pπ electron density into the empty 2p boron orbital. In <u>bis</u>aminoboranes, two nitrogen atoms compete for pπ-pπ overlap with the empty boron orbital, and we expect that each B-N linkage has a π-bond order (ideally) of at most 1/2. <u>Bis</u>aminoboranes should therefore have much lower B-N rotational barriers (10-11 kcal/mol). <u>Tris</u>aminoborane B-N rotational barriers should be lower still.

The value reported for compound <u>4</u>, then, indicates that only the B-NMe$_2$ bond has π character (*i.e.* this compound behaves as a <u>mono</u>aminoborane), and the bis(trimethylsilyl)amino B-N bond does not feature N$\rightarrow$B pπ-pπ overlap. Two arguments are developed to explain this and other data: steric arguments and competitive pπ-dπ bonding from N to Si, Ge, or Sn.

(b) Consider compounds such as <u>11</u>, <u>13</u>, and <u>15</u>. Also consider compounds <u>7</u>, <u>9</u>, and <u>14</u>. As the competition for the empty boron 2p orbital increases, the B-NMe$_2$ pπ-pπ bond should weaken. Where low B-NMe$_2$ rotational barriers are found, one can infer that the silylamino, germylamino or stannylamino N-B pπ-pπ bonding is extensive, and competes well with the NMe$_2$ group for the empty B orbital. Therefore, to the extent that the (Me$_3$E)RN$\rightarrow$B pπ-pπ bond is strong, the N$\rightarrow$EMe$_3$ pπ-dπ bond must be weak.

(c) If the silyamino nitrogen atom of compound <u>5</u> is oriented so that the nitrogen p atomic orbital is orthogonal to that of boron (the E$_2$N plane is perpendicular to the BNPh plane), then no pπ-pπ overlap between boron and the silylamino nitrogen can take place. The key question becomes--is the silyamino group rotated out of the plane for steric reasons or because there is no stabilizing E$_2$N$\rightarrow$B π bond? The latter result could arise because the π-electron density on the E$_2$N nitrogen prefers pπ-dπ bonding to silicon. Either argument (steric

or competetive π bonding) is consistent with the high B-NMe$_2$
rotational barrier. The authors have very nicely shown the
importance of steric factors by examining the compounds 7
through 10, and 14. N→Si pπ-dπ bonding is likely in compound
9. N→Sn pπ-dπ bonding is not likely in compound 14 because the Sn
4d orbital should have poor overlap with the N 2p orbital. The data
indicate a readiness in these compounds for E$_2$N-B pπ-pπ bonding
except where the size of the substituents prevents it.

15-3C

R$_2$SnX$_2$

Molecular association in the solid state leads to an increase in
apparent coordination number, eventually tending towards six-
coordinate pseudo-octahedral systems, although some Sn-X interactions
are only slightly closer than the sum of the van der Waals radii.
The interactions are described in a fashion analogous to the H-bond:
Y-A ··· X forms a three-centered, four-electron bond. Me$_2$SnF$_2$ is
associated extensively in the solid, forming F-bridged centers with
trans Me groups and four equivalent Sn-F bonds in a regular two-
dimensional array. In Me$_2$SnCl$_2$ and (CH$_2$Cl)$_2$SnCl$_2$, each Sn atom is
formally four-coordinate, but weak interactions in the solid from Sn
atoms to nearby Cl groups give weakly-linked chains of tetrahedra.
There seems to be some intermolecular association in crystals of
Ph$_2$SnCl$_2$ also. It is a general result, then, that association in the
solid increases the coordination number of Sn in its compounds with
the halides or psuedohalides. This is seen in the structures
reported here of Et$_2$SnCl$_2$ (Figure 1), Et$_2$SnBr$_2$ (Figure 2), and
Et$_2$SnI$_2$ (Figure 3).

SnCl$_4$(PR$_3$)$_2$

This is an unassociated molecular substance with a six-coordinate,
trans-octahedral structure. The Sn-Cl bond length (2.450 Å) is
longer than those of the above R$_2$SnX$_2$ compounds, confirming the trend
of increasing Sn-Cl bond length as the tendency towards octahedral
coordination increases.

15-5C

This is the red, tetragonal form of PbO. The oxygen atoms form the base of a square pyramid, and the lone pair of electrons is reasonably assigned to the position at the apex of each pyramid. The pyramids are oriented alternately up and down in layers, presumably due to the presence of the lone pairs.

15-7C

(a) This is molecule **1b**, which has its silicon-carbon framework in a single plane. Molecule **1a** is twisted, as shown in Figure 3 of the article.

(b) First there is a twist (angle 6.5°) along the silicon-silicon axis, as shown in Figure 3. Second there is what the authors have called a "pyramidalization" of the silicon atoms, as shown in Figure 2.

(c) The authors assign these to the steric effects of the bulky substituents. It is interesting, then, that compound **1b** does not exhibit these structural effects.

(d) The authors ascribe these differences to "lattice effects." However, there is also the discussion of electronic effects on the geometry of **1a**, which has two *cis* related mesityl rings that are coplanar with the Si=Si double bond and hence may be conjugated with this double bond. The geometry of **1b** is thought to be governed exclusively by steric factors.

A - Review

<u>16-1A</u>

The electron configuration of nitrogen is $1s^2 2s^2 2p^3$. Nitrogen fills its valence in its compounds in the following ways, with examples taken from the chapter:

$AB_x E_y$ Classification of Nitrogen	Hybridization at Nitrogen	Example
AB_4	sp^3	F_3B-NH_3, $NH_4{}^+$
		$[Co(NH_3)_6]^{3+}$
		$NF_4{}^+$
AB_3E	sp^3	NH_3, NF_3, NR_3
		N_2H_4, NH_2OH
AB_2E_2	sp^3	$NR_2{}^-$
AB_3	sp^2	$NO_3{}^-$, N_2O_4, CH_3NO_2
AB_2E	sp^2	$R_2C=N-OH$, XNO
		$R_2C=N-R$, $NO_2{}^-$
AB_2	sp	$NO_2{}^+$, $N_3{}^-$,
ABE	sp	N_2, RCN, NO^+

<u>16-3A</u>

This is the Ostwald process involving oxidation of ammonia:

$$2NH_3 + 5/2 O_2 \rightarrow 2NO + 3H_2O$$

followed by further oxidation, and then hydration:

$$NO + 1/2 O_2 \rightarrow NO_2$$
$$3NO_2 + H_2O \rightarrow 2HNO_3 + NO$$

16-5A

This is reaction 16-4.20, however note the competition from reaction 16-4.21.

16-7A

Ionizing solvents such as H_2SO_4 promote formation of the linear nitronium ion as in reaction 16-6.1 and 16-6.2, or as in reactions 16-6.4 and 16-6.5. The attacking reagent in aromatic nitrations is the nitronium ion, as shown in reaction 16-6.3.

16-9A

Nitrite ligands can be either N-bound or O-bound, and nitro/nitrito isomerism was discussed in Chapter 6. Nitrate acts as a ligand in numerous ways: monodentate ($M-ONO_2$), chelating (MO_2NO), bridging [M-ON(O)O-M], or triply bridging [M-ON(O-M)O-M].

16-11A

These molecules have odd electrons, and they readily dimerize to form N-N bonds:

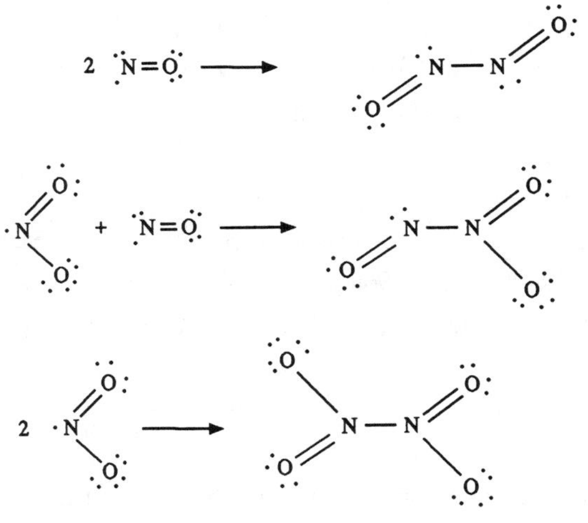

B - Additional Exercises

16-1B

The molecular orbital theory of diatomic molecules, as developed in Section 3-5 (especially Figures 3-26 and 3-27), is appropriate for all of these isoelectronic molecules and ions. The MO electron configuration for each of these is comparable to that of either Figure 3-27a or 3-27b. Each molecule or ion has a triple bond. CO and CN⁻ very readily form complexes with metals, especially those in low oxidation states. CO more readily forms complexes than N_2 because, as discussed in Chapter 28, N_2 is both a weaker σ-donor and a weaker π-acceptor.

16-3B

Using the data of Table 1-1, we can see that NF_3 is formed exothermically from the elements:

$$1/2N_2 + 3/2F_2 = NF_3$$
$$\Delta H = -[3 \times E_{N-F} - (1/2 \times E_{N \equiv N} + 3/2 \times E_{F-F})]$$
$$\Delta H = -3 \times 272 + 1/2 \times 946 + 3/2 \times 158$$
$$\Delta H = -106 \text{ kJ}$$

whereas NCl_3 is formed endothermically:

$$1/2N_2 + 3/2Cl_2 = NCl_3$$
$$\Delta H = -[3 \times E_{N-Cl} - (1/2 \times E_{N \equiv N} + 3/2 \times E_{Cl-Cl})]$$
$$\Delta H = -3 \times 193 + 1/2 \times 946 + 3/2 \times 242$$
$$\Delta H = +257 \text{ kJ}$$

The distinguishing differences between these two systems are (1) the great strength of the N-F bond compared with the relative weakness of the N-Cl bond, and (2) the weakness of the F-F bond compared with the Cl-Cl bond.

16-5B

The reactions are reproduced here, with the oxidation state of nitrogen in each compound listed beneath the equations:

$$2HNO_3 = NO_2^+ + NO_3^- + H_2O \qquad (16\text{-}6.1)$$

V   V   V

$$N_2O_4 = NO^+ + NO_3^- \qquad (16\text{-}5.17)$$
$$\text{IV} \quad \text{III} \quad \text{V}$$

$$2NO_2 = 2NO + O_2 \qquad (16\text{-}5.14)$$
$$\text{IV} \qquad \text{II}$$

$$3\ NO = N_2O + NO_2 \qquad (16\text{-}5.7)$$
$$\text{II} \quad \text{I} \qquad \text{IV}$$

$$.\ NH_4^+NO_3^- = N_2O + 2H_2O \qquad (16\text{-}5.1)$$
$$\text{-III} \quad \text{V} \qquad \text{I}$$

$$NH_4^+NO_3^- = NH_3 + HNO_3 \qquad (16\text{-}4.13)$$
$$\text{-III} \quad \text{V} \qquad \text{-III} \qquad \text{V}$$

$$4NH_3 + 5O_2 = 4NO + 6H_2O \qquad (16\text{-}4.4)$$
$$\text{-III} \qquad \qquad \text{II}$$

$$N_2 + O_2 = 2NO \qquad (16\text{-}2.4)$$
$$\text{O} \qquad \qquad \text{II}$$

Of these equations, two are not redox reactions (16-6.1 and 16-4.13). The other reactions are redox reactions. Additionally, 16-5.1, 16-5.7, and 16-5.17 are disproportionation reactions.

16-7B

The lone pairs are shown below. The σ bonds have been listed simply as lines, for the sake of clarity. The hybridization of each lone-pair orbital is noted.

(a) N_2

(b) N_3^-

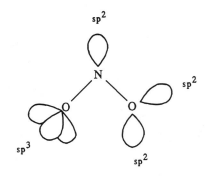

(c) NO_2^-

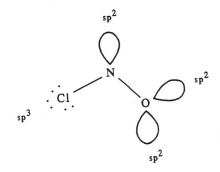

(d) ClNO

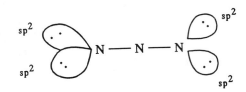

16-9B

(a) $3Li + 1/2N_2 \rightarrow Li_3N$

(b) $2Cu + NO_2 \rightarrow Cu_2O + NO$

(c) $C + NO_2 \rightarrow CO_2 + 1/2N_2$

(d) $2NO_2 + H_2O_2 \rightarrow 2HNO_3$

(e) $2NO_2 + O_3 \rightarrow N_2O_5 + O_2$

(f) $2H_2 + NO_2 \rightarrow 1/2N_2 + 2H_2O$

  $7H_2 + 2NO_2 \rightarrow 2NH_3 + 4H_2O$

(g) $2HI + 2HNO_2 \rightarrow I_2 + 2NO + 2H_2O$

16-11B

Oxide	AB_xE_y Classification at Nitrogen
N_2O	ABE_2, end N
	AB_2, middle N
NO (a radical)	$ABE_{1.5}$
NO_2 (a radical)	$AB_2E_{0.5}$
N_2O_3	AB_2E, the -NO nitrogen
	AB_3, the $-NO_2$ nitrogen
N_2O_4	AB_3

16-13B
See Figures 16-2 and 16-3 and Tables 16-1 and 16-2.

(a) $2NO_2 + H_2O \rightarrow HNO_3 + HNO_2$
    $2Cu + NO_2 \rightarrow Cu_2O + NO$

There is also the route of stepwise disproportionation of NO_2:

$$2NO_2 + H_2O \rightarrow HNO_3 + HNO_2$$
$$3HNO_2 \rightarrow HNO_3 + 2NO + H_2O$$

(b) Two paths give us N_2 from NO_2:

$$C + NO_2 \rightarrow CO_2 + 1/2N_2$$
$$2H_2 + NO_2 \rightarrow 1/2N_2 + 2H_2O$$

$$N_2 + 3H_2 \rightarrow 2NH_3$$

16-15B
This is essentially equation 16-4.7.

C - Questions from the Literature of Inorganic Chemistry

16-1C

(a)

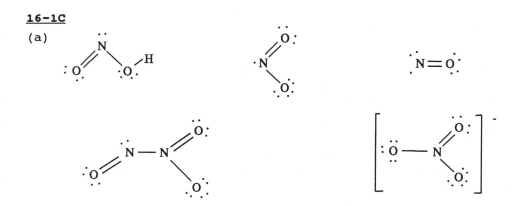

In every case the hybridization at nitrogen is sp^2. The oxidation state for oxygen is taken to be -2 and that for hydrogen is +1. The oxidation states for nitrogen are then:

HNO_2, +3.
N_2O_3, +2 and +4, or +3 each.
NO, +2.
NO_2, +4.
NO_3^-, +5.

(b) The principal evidence for the production of N_2O_3 is spectroscopic. First, there is the broad and weak absorbance ($\lambda_{max} \sim 650$ nm, $\varepsilon_{max} \sim 20$ M^{-1} cm^{-1}) that is responsible for the blue color of HNO_2 in concentrated sulfuric and perchloric acids. Secondly, there is a more pronounced absorption band at 220 nm that was used to advantage in this study. Note Figure 3 especially.

(c) Equation (1) is the sum of equations (3), (13), (14), and (2), with coefficients in equations (3) and (13) changed so that the stoichiometry of the sum is correct. The equilibrium constant for equation (1) is then the product of the equilibrium constants for the individual steps.

(d) This was calculated as follows. The value of K_1 allows

$\Delta G°(1)$ for reaction (1) to be calculated:

$$\Delta G°(1) = -RT \ln K_1$$

The remaining values for equation (1) then are:

$$\Delta G°(1) = \Delta G°_f[N_2O_3(aq)] + \Delta G°_f[H_2O(l)] - \Delta G°_f[HNO_2]$$

from which it can be calculated that $\Delta G°_f[N_2O_3(aq)] = 33.5$ kcal/mol.

(e) Consider the oxidation states that were assigned in part (a) of this question. The reactions may be classified as follows.

      disproportionations: (3) and the reverse of (13)
                hydrolysis: (1) and the reverse of (13)
                  acid-base: (1), (2), and (9)

16-3C

(a) (i) $NF_4ClO_4 \rightarrow NF_3 + FOClO_3$
   (ii) $NF_4BrO_4 \rightarrow NF_3 + FBrO_2 + O_2$
  (iii) $NF_4HF_2 \cdot nHF \rightarrow NF_3 + F_2 + (n+1)HF$

(b) (i) $NF_4^+ + BrF_4^- \rightarrow NF_3 + BrF_5$
   (ii) $NF_4^+ + BrF_4O^- \rightarrow BrF_3O + F_2 + NF_3$

A - Review

17-1A

Nitrogen forms strong $p\pi$-$p\pi$ bonds, especially the triple bond in N_2, which accounts for the inertness of elemental nitrogen. Conversely, in compounds of nitrogen, the N-N single bond energy is low. It is, in particular, quite a bit lower than the P-P single bond energy. (Recall the data of Table 1-1.) The great strength of the P-P single bond favors catenation in the element (P_4).

17-3A

These are comparable to the factors that cause a similar difference between the elements C and Si, namely the inability of phosphorus (Si) to form $p\pi$-$p\pi$ bonds and the ability of phosphorus (Si) to form $p\pi$-$d\pi$ bonds. Also, the phosphorus atom is larger and has a lower ionization potential than nitrogen.

17-5A

(i) $P_4 + 10HNO_3 + H_2O \rightarrow 4H_3PO_4 + 5NO + 5NO_2$

(ii) $AsCl_3 + 3H_2O \rightarrow H_3AsO_3 + 3HCl$

(iii) $OPCl_3 + 3H_2O \rightarrow H_3PO_4 + 3HCl$

(iv) $3P_4O_{10} + 12HNO_3 \rightarrow 4(HPO_3)_3 + 6N_2O_5$
Note: $(HPO_3)_3$ is the cyclic trimer, *cyclo*-trimetaphosphoric acid, consisting of a six-membered ring of alternating P and O atoms with exocyclic (=O) and (-OH) groups at each phosphorus atom.

(v) $P_4O_6 + 6H_2O \rightarrow 4H_3PO_3$

(vi) $Zn_3P_2 + 6HCl \rightarrow 2PH_3 + 3ZnCl_2$

17-7A

The structure of P_4O_{10} is shown in Figure 17-2a. As_4O_6 has a structure that is similar to that of P_4O_6 (Figure 17-2b), as does Sb_4O_6.

17-9A

(a) Phosphorus acid is shown in Structure 17-I. The sp^3-hybridized phosphorus atom is of the AB_4 classification.

(b) Triethylphosphite is $P(OEt)_3$. It has a pyramidal structure. The sp^3-hybridized phosphorus atom is of the AB_3E classification.

17-11A

Extensive hydrogen-bonding between molecules of H_3PO_4 gives the pure liquid a high viscosity.

17-13A

The phosphazenes are ring or chain compounds of alternating N and P atoms, with two additional substituents at each phosphorus. Structures 17-III through 17-VIII are examples. They are made as in reaction 17-9.1.

17-15A

PF_5 is prepared by fluorination of PCl_5 with CaF_2 at 400 °C:

$$2PCl_5 + 5CaF_2 \rightarrow 2PF_5 + 5CaCl_2$$

PCl_5 has a trigonal bipyramidal geometry. It readily adds an electron pair donor to give an octahedral adduct, an example being $[PF_6]^-$.

B - Additional Exercises

17-1B

Three principal effects arise from the importance of d-orbital participation in the bonding of phosphorus. First, even in compounds that are electronically saturated as defined in Chapter 3, bonds to phosphorus, especially those involving oxygen and nitrogen, are often shorter than would be expected for a purely single bond. This implies some degree of multiple bonding of the $d\pi$-$p\pi$ type. As an example consider the greater strength of the P=O bond in phosphine oxides, compared with the weaker N-O bond of corresponding amine oxides. Second, the use of d-orbitals allows for coordination number expansion at phosphorus through the use of dsp^3 and d^2sp^3 hybridizations. Thus phosphorus compounds with occupancies (the sum $x + y$ in systems AB_xE_y, as defined in Chapter 3) at phosphorus of five and six are common, examples being PF_5 (an AB_5 system) and $[PF_6]^-$ (an AB_6 system). Occupancies greater than four are not observed for nitrogen. Third, AB_3E systems such as PR_3 and PX_3 generally form stronger bonds to transition metals and the like, because of the possibility for $d\pi$-$d\pi$ bonding, in which the phosphorus atom serves as a π acid.

17-3B

PCl_5 and $SbCl_5$ have been known for some time, but $AsCl_5$ was prepared only recently by K. Seppelt. (See *Angew. Chem., International Edition in English,* **1976**, *15*, 377-8.) It is made at -105 °C by oxidation of $AsCl_3$ in chlorine. $AsCl_5$ decomposes at -50 °C to $AsCl_3$ and Cl_2. $BiCl_5$ is still not known. No completely satisfying explanation for the instabilities of $AsCl_5$ or $BiCl_5$ has been made, but two factors seem to be important. First, the effective nuclear charge on the As^V and Bi^V ions is very high, and promotion to give dsp^3 hybridization is consequently unfavorable. Second, these oxidation states are highly oxidizing. This accounts especially for the nonexistence of the five-coordinate pentabromides and pentaiodides of the Group VA(15) elements, since bromide and iodide are readily reduced.

17-5B

These are $d\pi$-$p\pi$ bonds analogous to that shown in Figure 15-1 for the silyl amines:

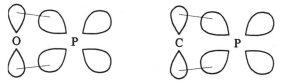

Each phosphorus atom falls into the AB_4 classification, is sp^3-hybridized, and is tetrahedrally coordinated.

17-7B

(a) $PCl_3 + 3H_2O \rightarrow H_3PO_3 + 3HCl$

(b) $PCl_3 + 1/2O_2 \rightarrow Cl_3PO$

(c) $BiCl_3 + H_2O = BiOCl + 2HCl$

(d) See Figure 17-1 and reaction 29-10.1:

$PCl_3 + 3EtMgBr \rightarrow PEt_3 + 3MgBrCl$

(e) $PCl_3 + F_2 \rightarrow PCl_3F_2$

(f) by analogy to hydrolysis, and as shown in Figure 17-1:

$Cl_3PO + 3MeOH \rightarrow (MeO)_3PO + 3HCl$

(g) $2PCl_5 \rightarrow [PCl_4]^+ + [PCl_6]^-$

(h) $PCl_3 + 3NH_3 \rightarrow P(NH_2)_3 + 3HCl$

(i) $3PCl_5 + 3NH_4Cl \rightarrow (NPCl_2)_3 + 12HCl$

17-9B

A number of conformations are possible, but only three geometrical isomers exist.

The geminal isomer:

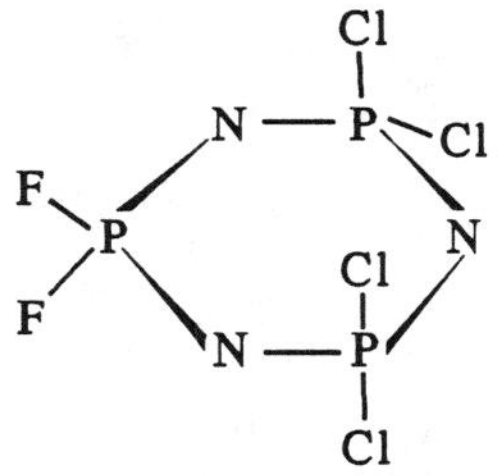

The nongeminal, trans isomer:

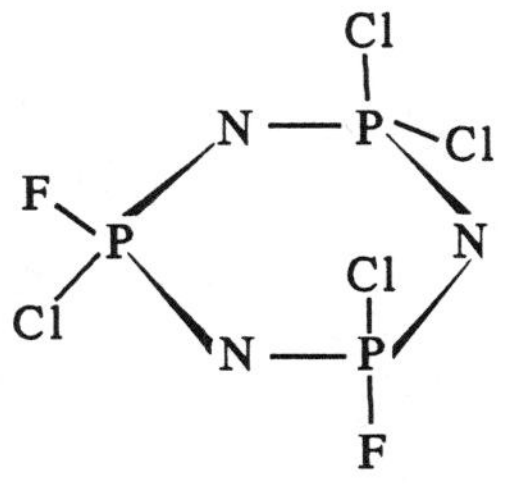

The nongeminal, cis isomer:

Lone pairs have been omitted for clarity.

17-11B

(a) SbF_5 is a Lewis acid, reacting with the Lewis base F^- to give $[SbF_6]^-$.

(b) PCl_5 exchanges Cl^- with another PCl_5 to form the ionic compound $[PCl_4][PCl_6]$. Thus one PCl_5 acts as a Lewis acid, accepting a pair of electrons from Cl^-.

(c) This is similar to (a) above.

(d) SbF_5 accepts a pair of electrons and, as such, could be considered to behave as a Lewis acid. Technically, this is simple reduction.

17-13B

(a) This is equation 17-2.1.

(c) $OPCl_3 + 3H_2O \rightarrow H_3PO_4 + 3HCl$

(c) $PCl_3 + 3PhOH \rightarrow P(OPh)_3 + 3HCl$

(d) $P_4 + 5O_2 \rightarrow P_4O_{10}$

(e) $P(OPh)_3 + 1/2O_2 \rightarrow OP(OPh)_3$

(f) $PCl_3 + 3EtMgBr \rightarrow PEt_3 + 3"MgBrCl"$

(g) $PCl_3 + 3MeOH \rightarrow P(OMe)_3 + 3HCl$

(h) $PCl_3 + AsF_3 \rightarrow PF_3 + AsCl_3$

(i) $4PCl_5 + 10H_2 \rightarrow P_4 + 20HCl$

17-15B

Even in a molecule that is electronically saturated, i.e. $O=PCl_3$ in which the phosphorus atom is sp^3-hybridized, there may also be $d\pi-p\pi$ bonding, which shortens the phosporus-element bond.

17-17B

Sb_2O_3 is amphoteric. It is the oxide of a metalloid.

17-19B

(a) $[PCl_4]^+[NbCl_6]^-$

(b) $[PCl_4]^+[PCl_6]^-$

(c) PF_5

(d) $[NPCl_2]_n$

(e) $[NP(OEt)_2]_3$

(f) $[NP(Ph)_2]_3$

(g) $[NP(Ph)_2]_3$

(h) PPh_3

(i) $[PCl_4]^+[Ti_2Cl_9]^-$ plus $[PCl_4^+]_2[Ti_2Cl_{10}{}^{2-}]$

17-21B

See Section 17-11.

C - Questions from the Literature of Inorganic Chemistry

17-1C

(a) The anions $[SbF_6]^-$ and $[AsF_6]^-$ have the same octahedral structure as $[PF_6]^-$:

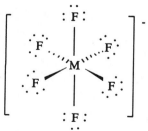

The cations $[As_3S_4]^+$ and $[As_3Se_4]^+$ are isostructural. They are cage compounds with $-S-$ or $-Se-$ bridging on three out of the six edges

formed by the tetrahedron of three As atoms and one S atom (in
$[As_3S_4]^+$) or the tetrahedron of three As atoms and one Se atom (in
$[As_3Se_4]^+$):

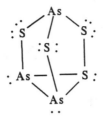

There are thirty-eight electrons distributed into the above Lewis
diagram. This is also the structure of the P_4S_3 compound mentioned
in Problem 17-8A.

(b) $SbCl_5$ reacts with As_4S_4 to give $SbCl_3$, $AsCl_3$, and sulfur. Cl_2
reacts with As_4S_4 to give $AsCl_3$, S_2Cl_2, and sulfur. This occurs
because the reagents are more powerful oxidizing and halogenation
reagents. For instance, PCl_5 does not oxidize As_4S_4 in methylene
chloride.

(c) As shown in Table III, both are cage compounds, $[As_3S_4]^+$ being
one vertex smaller than As_4S_4. There is an extended discussion of
As-As and Sb-Sb bond lengths in the article. It is instructive to
compare the structure of these sulfides with the oxides of phosphorus
shown in Figure 17-2 of the text.

17-3C

(a)

Compound	AB_xE_y Classification of Sb	Hybridization at Sb	Geometry at Sb
$SbCl_3$	AB_3E	sp^3	pyramidal
$(NH_4)_2SbCl_5$	AB_5E	d^2sp^3	square-pyramidal
$[Co(NH_3)_6][SbCl_6]$	AB_6	d^2sp^3	octahedral

The structure of [pyH][SbCl$_4$] involves extensive sharing of bridging Cl atoms between SbCl$_4$ units, giving very distorted geometry about each antimony in the solid.

(b) If the lone pair electron density were directed along the axis of the sixth octahedral coordination site on each [SbCl$_5$]$^{2-}$, then the structure with two base-to-base pyramids would be destabilized by electron-electron repulsion between the lone pairs of the two adjacent Sb^{III} centers. Since the base-to-base structure is nevertheless observed, the lone pair electrons must not have much of a "stereochemical effect."

A - Review

18-1A

$1s^2 2s^2 2p^4$

18-3A

The carbon-oxygen bond in acetone is a double bond. It has two components: a σ bond formed by the overlap of an sp^2-hybrid orbital on carbon with an sp^2 hybrid orbital on oxygen. The second of the two bonds from carbon to oxygen is a π bond, formed by the side-to-side overlap of unhybridized p-atomic orbitals of carbon and oxygen. The C-O σ bond has no nodal plane that includes the C-O internuclear axis, whereas the π bond has one nodal plane that includes the C-O internuclear axis.

18-5A

As shown in Figure 3-26, the dioxygen molecule has two unpaired electrons in its MO electron configuration:

$$[\sigma_1]^2 [\sigma_2]^2 [\sigma_3]^2 [\pi_1]^4 [\pi_2]^2$$

18-7A

Ozone may be prepared in the laboratory by processes that cleave oxygen molecules: electric discharge or UV irradiation.

18-9A

(a) $6H^+ + 2MnO_4^- + 5H_2O_2 \rightarrow 2Mn^{2+} + 5O_2 + 8H_2O$

(b) Recall equations 5-2.19 through 5-2.22:
$8OH^- + 4Fe^{2+} + O_2 \rightarrow 2Fe_2O_3 \cdot 3H_2O(s) + H_2O$

(c) $2CO_2 + 2Na_2O_2 \rightarrow 2Na_2CO_3 + O_2$

(d) $2O_2^- + H_2O \rightarrow O_2 + HO_2^- + OH^-$

B - Additional Exercises

18-1B

See the answer to question 3-1B. The O_2^+ cation, having one unpaired electron, should have a magnetic moment of 1.73 BM.

18-3B

(a) In ketones:

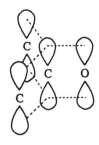

(b) In carbonate:

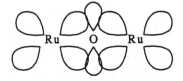

(c) In $[Cl_5Ru-O-RuCl_5]^{4-}$:

(d) In ozone:

(e) In triphenylphosphine oxide:

(f) In disiloxanes:

(g) In $OSCl_2$:

C - Questions from the Literature of Inorganic Chemistry

18-1C

A great deal of progress has been made in understanding the reversible bonding of dioxygen to metal complexes. The student is encouraged to compare these two historically important examples of successfully reversible "synthetic oxygen carriers" with those discussed in Chapter 31, and the supplementary readings listed there. See especially the extensive review by R. D. Jones, D. A. Summerville and F. Basolo, "Synthetic Oxygen Carriers Related to Biological Systems," in *Chemical Reviews*, **1979**, *79*, 139-179

(a) Consider first the reversible adduct of Crumbliss and Basolo. The magnetic susceptibility data suggest a single unpaired electron. The paper that follows this one makes it clear that the electron should be considered to reside on the dioxygen ligand. Therefore, of the two possible formulations ($Co^{II}-O_2$ vs. $Co^{III}-O_2^-$), the second is most reasonable. The infrared data indicate a coordination geometry for the superoxide ligand as shown in Figure 18-2a of the text, corresponding to structure IV in the article by Crumbliss and Basolo.

Next consider the O_2-adduct of Vaska's compound as studied by La Placa and Ibers. The O-O distance reported in this article (1.30 Å) indicates that the dioxygen ligand is most similar to ionic superoxides, O_2^-, for which the O-O distance is equal to 1.28 Å. (The O-O distance in ionic peroxides is typically 1.49 Å, whereas that in dioxygen is 1.21 Å.)

(b) If the O_2-adduct of Vaska's compound were to be formulated as containing a superoxide ligand, we would write $[Ir^{II}(O_2)(Cl)(P(C_6H_5)_3)_2]$, and it might be reasonable to expect the compound to be paramagnetic. If we consider the dioxygen ligand to be a peroxide, we can simply write $[Ir^{III}(O_2)(Cl)(P(C_6H_5)_3)_2]$, and both the O_2^{2-} ligand and the d^6 Ir^{III} center are straightforwardly diamagnetic.

These simplistic approaches are initially useful, but more delocalized molecular orbital studies have proved to be more successful in understanding the reversibility, magnetism, and coordination geometry of dioxygen adducts. More is given in Chapter 31, and the student is encouraged to note the articles listed in the supplementary readings for Chapters 18 and 31, and the review given above. The most obvious difference between molecular oxygen and one of its adducts is that the $\pi*$ levels that are doubly degenerate in dioxygen need not stay doubly degenerate upon complexation to a metal. Thus a molecule of O_2 that is coordinated end-on should be diamagnetic because the degeneracy of the $\pi*$ orbitals is lifted by the bent geometry.

(c) The data for dioxygen O-O distances are: O_2 (1.21 Å), O_2^- (1.28 Å), and O_2^{2-} (1.49 Å). A value in the range 1.26 to 1.30 Å should be reasonable for the compounds of Basolo and Crumbliss. Some interesting data are available in a paper by F. Basolo, B. M. Hoffman and J. A. Ibers, *Accounts of Chemical Research*, **1975**, *8*, 384-392. The O-O distance in a closely similar adduct has been reported (Rodley, G. A. and Robinson, W. T., *Nature (London)*, **1972**, *235*, 438) to be 1.26 ± 0.04 Å.

A - Review

19-1A

Sulfur occurs in nature mostly as the element and as sulfide and sulfate ores. Additionally, there are H_2S deposits in natural gas, and atmospheric SO_2.

19-3A

In superacids such as oleum, di-cations such as those of Structure 19-I and 19-II are obtained.

19-5A

First, we have the tetrafluoride, SF_4, which is exceptionally sensitive to hydrolysis. In contrast, the hexafluoride, SF_6, is quite inert. Sulfur in SF_6 is octahedral (AB_6 classification), whereas in SF_4 the sulfur is of the AB_4E classification. Last we have the dimeric S_2F_{10}, which contains a weak S-S bond, and a staggered arrangement of fluorine atoms. Like SF_6, S_2F_{10} is unreactive, although it does disproportionate at 150 °C to give SF_4 and SF_6.

19-7A

The most important reaction of SO_3 is with water:

$$SO_3 + H_2O \rightarrow H_2SO_4$$

After that we have various sulfonations and oxidations.

19-9A

Sulfuric and selenic acid are similar, the latter being a better oxidizing agent. Telluric acid is quite different; it is composed of octahedral $Te(OH)_6$ molecules in the solid. In fact, most tellurates are octahedral, as in the anion $[H_5TeO_6]^-$, which is properly written $[TeO(OH)_5]^-$.

CHAPTER 19

B - Additional Exercises

19-1B

The boiling points of the hydrides H_2X are compared in Figure 9-1.
Except for H_2O, which is out of place due to intermolecular hydrogen
bonding, the boiling points increase from top to bottom in the
group. This is because the London dispersion forces are stronger in
the larger, more polarizable molecules. The acidity of the Group
VIB(16) hydrides increases from top to bottom in the group. This is
because the X-H bond strength decreases from top to bottom in the
group.

19-3B

SF_6 may be made by direct reaction of S_8 with fluorine. SF_4 is
prepared as in reaction 19-4.1. SF_4 is a useful and selective
fluorinating agent as outlined in Section 19-4. SF_6 is sufficiently
inert to be used as an insulating agent in electrical devices such as
high-voltage generators.

19-5B

This compound may be regarded as a diadduct of $SeOCl_2$ and two
pyridine molecules. The selenium atom in the diadduct is of the AB_5E
classification, and the geometry is square pyramidal:

19-7B

Molecule or Ion	AB_xE_y Classification of Sulfur	Hybridization at Sulfur	Geometry at Sulfur
$S_2O_3^{2-}$	AB_4 (central atom)	sp^3	tetrahedral
$S_2O_4^{2-}$	AB_3E	sp^3	pyramidal
$S_2O_6^{2-}$	AB_4	sp^3	tetrahedral
SO_2	AB_2E	sp^2	bent
SO_3	AB_3	sp^2	triangular
$SOCl_2$	AB_3E	sp^3	pyramidal
SO_2Cl_2	AB_4	sp^3	tetrahedral
SCl_2	AB_2E_2	sp^3	bent
S_2Cl_2	AB_2E_2	sp^3	bent

19-9B

The adduct between $SbCl_5$ and Cl_3PO has the following geometry:

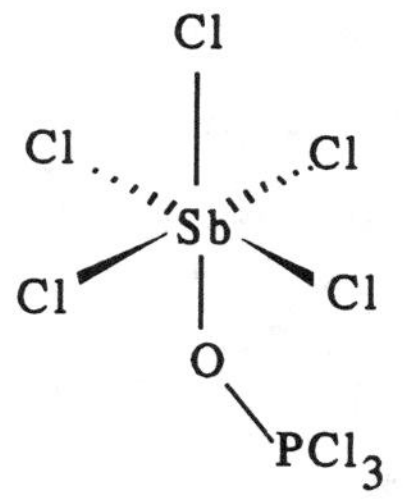

The Sb–O–P angle is 146°.

19-11B

First we should point out that there is a $p\pi$-$p\pi$ bond exactly as described in the general AB_3 system of Chapter 3, Figures 3-30, 3-31, and 3-32. If we define the molecular plane to coincide with the xy plane, then $d\pi \leftarrow p\pi$ overlap is also possible both with the d_{xz} and the d_{yz} orbitals of sulfur. The three Group Orbitals GO_1, GO_2 and GO_3 may overlap with the S-centered d_{xz} and d_{yz} orbitals as follows.

bonding overlap with GO_1:

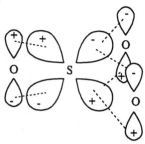

nonbonding overlap with GO_3 and GO_2:

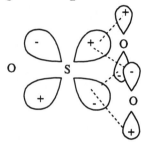

antibonding overlap with the negative of GO_1:

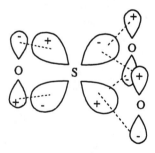

19-13B

Any bond length shorter than the sum of the single bond covalent radii (available in Figure 2-15) implies some degree of multiple bonding. The amount of π bonding is taken to occur in proportion to the amount by which bond lengths are shorter than these single bond standards. These single bond standards are S-O (1.73 Å) and S-S (2.06 Å). (S-O bond distances are unavailable for $S_3O_6^{2-}$ and $S_4O_6^{2-}$.) Of the remaining substances, the π-bond system is greatest in SO_3 (because this molecule has the shortest S-O distance), followed in decreasing π-bond order by SO_4^{2-}, $S_2O_3^{2-}$, and $S_2O_4^{2-}$. It is assumed in making this comparison that neutral molecules and dianions can be compared. The S-S bond in $S_2O_3^{2-}$ and one of the S-S bonds in $S_4O_6^{2-}$ are short enough to suggest some degree of π-bonding in these ions. The other S-S bonds in this series are longer and weaker than the typical S-S single bond.

C - Questions from the Literature of Inorganic Chemistry

19-1C

(a) The Lewis diagrams are the following, and a table of other properties is presented on the next page.

SeO_2F_2

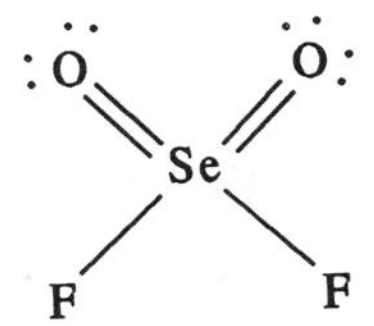

$SeOF_4$

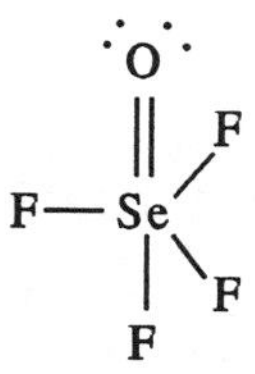

$Se_2O_2F_8$

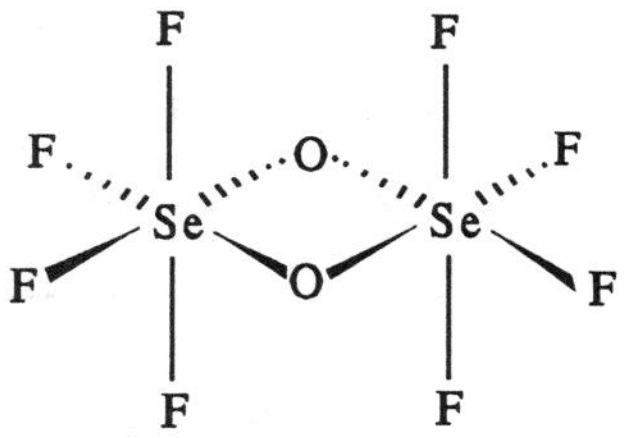

$Te_2O_2F_8$

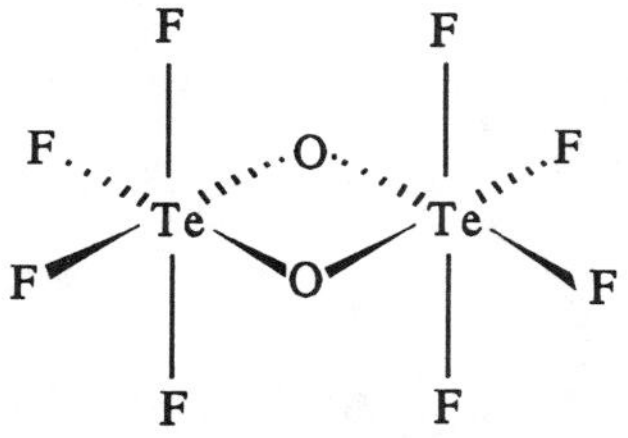

$I_2O_4F_6$

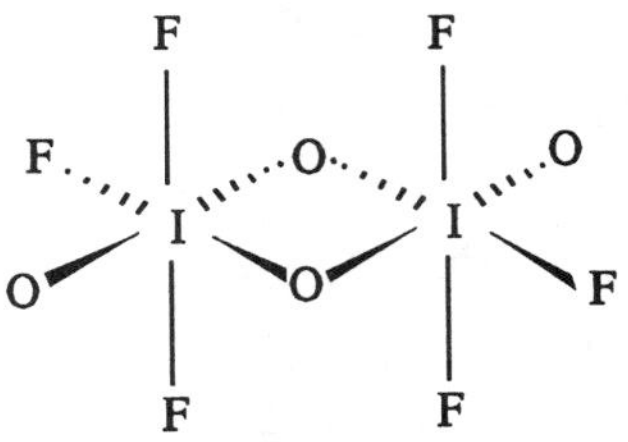

$F_5S-O-SF_5$

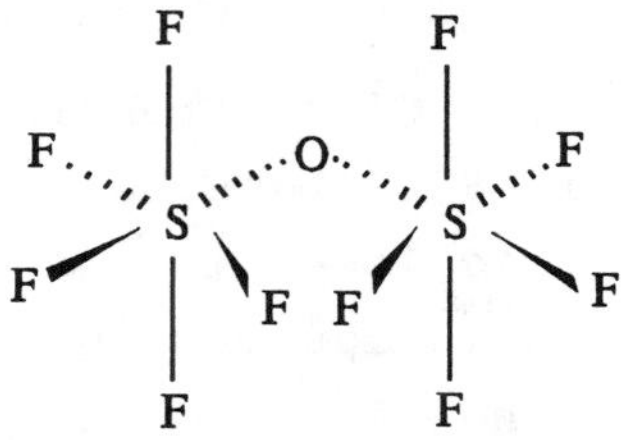

$F_5Se-O-SeF_5$

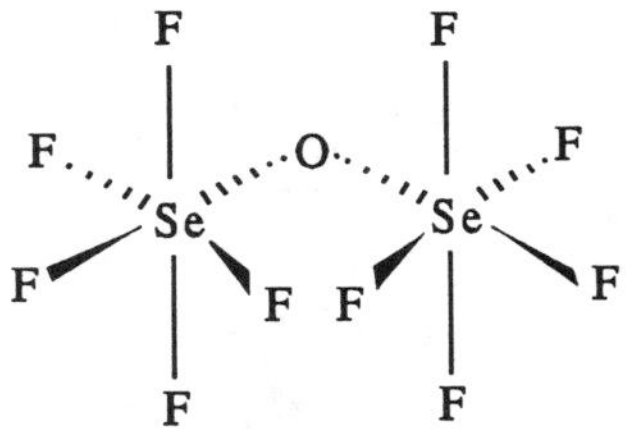

Note: Three lone pairs on each fluorine atom have been omitted for clarity.

Molecule	AB_xE_y Classification	Hybridization	Geometry
SeO_2F_2	Se – AB_4	sp^3	tetrahedral
$SeOF_4$	Se – AB_5	dsp^3	trigonal-bipyramid
$Se_2O_2F_8$	Se – AB_6	d^2sp^3	pseudo-octahedral
$Te_2O_2F_8$	Te – AB_6	d^2sp^3	pseudo-octahedral
$I_2O_4F_6$	I – AB_6	d^2sp^3	pseudo-octahedral
$F_5S-O-SF_5$	S – AB_6	d^2sp^3	pseudo-octahedral
$F_5Se-O-SeF_5$	Se – AB_6	d^2sp^3	pseudo-octahedral

The geometries that are listed in the above table are the first approximation geometries at the "central" X atom. Geometries are bent at the bridging oxygens, and these four-membered ring-dimers have strained structures.

(b) There is substantial π bonding between Se and O in both monomers, SeO_2F_2 and $SeOF_4$. The structures with four-membered rings (i.e. the dimers) are assumed to have little X-O π component, inspite of the fact that the X-O distance in the bridges is somewhat shorter than expected for simple single bonds. Some Se-Se bonding or Te-Te bonding (presumably by $d\pi$-$d\pi$ overlap) in the dimers is invoked. Note also the last sentence under Discussion. There is evidently some π component to the X-O bond in the substances F_5X-O-XF_5 (X = Se, Te) as indicated in Table IV by the short X-O distances in these compounds.

(c) $SeOF_4$ dimerizes, according to these authors, in order to avoid Se-O double bonds in the monomers.

(d) $TeOF_4$ has not been isolated at the time of publication, although the dimer is apparently stable.

19-3C

(a) The best synthesis begins with reaction of selenium with chlorine trifluoride:

$$3Se + 4ClF_3 \rightarrow 3SeF_4 + 2Cl_2$$

Next, the Cs^+ salt of $[SeF_5]^-$ was prepared:

$$SeF_4 + CsF \rightarrow CsSeF_5$$

The last step is chlorination using $ClSO_3F$:

$$CsSeF_5 + ClSO_3F \rightarrow CsSO_3F + SeF_5Cl$$

The last reaction may also be considered to be a metathesis reaction. Another (less efficient) synthesis employs the method that had worked successfully for SF_5Cl:

$$SeF_4 + ClF \rightarrow SeF_5Cl$$

The last method, although direct, gave numerous, unwanted side products.

(b) Hydrolysis affords titratable F^- and selenate, SeO_4^{2-}:

$$SeF_5Cl + 8OH^- \rightarrow SeO_4^{2-} + 4H_2O + 5F^- + Cl^-$$

(c) The series is SF_5Cl, SeF_5Cl, and TeF_5Cl. The infrared data suggest a similar structure for these molecules, as do the NMR spectra.

A - Review

20-1A

The halogens generally occur in nature as halides in minerals, salt deposits, brines and the like. Additionally, iodine can be found as the iodate, for example $Ca(IO_3)_2$ and $7Ca(IO_3)_2 \cdot 8CaCrO_4$, which are recoverable for Chilean nitrate deposits. None of the halogens occurs naturally in the uncombined, elemental form. Examples of the various minerals and ores for each of the halogens are listed in Section 20-2.

20-3A

Anhydrous compounds of the halides with other elements may be prepared by (a) direct interaction of the element with one of the halogens, or of many metals with the hydrohalic acids HX, (b) dehydration of hydrated halides, although oxohalides sometimes result, (c) displacement of oxide by halide, as in reactions 20-3.3 through 20-3.5, and (d) exchange of one halogen for another (especially of iodide and bromide for chloride) by metathesis.

20-5A

The iodide ion is a good reducing agent. Consider, for example, the following reaction:

$$Cu^{2+} + 2I^- \rightarrow CuI + 1/2 I_2$$

20-7A

Bridging halogen atoms are common among the so-called molecular halides mentioned in Section 20-3. One, two, and three bridging halogen atoms between pairs of metal atoms are known. Most commonly we have structures 20-I, 20-II and 20-III. Bromide and chloride bridges are characteristically bent, whereas fluoride bridges may also be linear, as in structure 20-III.

20-9A

The highest oxidation state for the halogens is represented by the perhalic acids (HXO_4) and the corresponding perhalates (XO_4^-). In descending order of oxidation state we have also halic acids (HXO_3) and halates (XO_3^-), halous acids (HXO_2) and halites (XO_2^-), and finally hypohalous acids (HXO) and hypohalites (XO^-). Uniquely for iodine, there is also ortho-periodic acid, H_5IO_6, which should be written $(HO)_5IO$ (*cf.* telluric acid).

20-11A

The soluble triiodide ion is readily formed in aqueous solutions:

$$I_2 + I^- \rightarrow I_3^-$$

B - Additional Exercises

20-1B

The O-O distance in O_2F_2 is comparable to that in dioxygen (1.21 Å), which has an O=O double bond. A single bond between oxygen atoms has a length of *ca.* 1.49 Å, as found in H_2O_2 or O_2^{2-}. There is no satisfying resonance explanation for the short O-O bond in O_2F_2. Molecular orbital approaches, frankly, are not straightforward. Perhaps a three-center, four-electron bond covering two separate OOF units is plausible.

20-3B

If we consider only the unassociated molecule SbF_3, we predict antimony to fall in the AB_3E classification, and we expect the molecule to have pyramidal geometry. In the solid, there is some tendency for association through fluorine-bridging, although the geometry around Sb is nearly that of a regular pyramid. In solid BiF_3, intermolecular association *via* fluorine-bridging gives Bi an effective coordination number of eight. Such bridging is a typical structural feature of the molecular halides, as discussed in Section 20-3. In the solid state, $SbCl_5$ has a discrete, molecular, trigonal bipyramidal structure typical of an AB_5 system, and there is no Cl-bridging between one Sb and another.

20-5B

The acid strengths are $HClO_4 > HClO_3 > HClO_2 > HClO$. In this series of oxyacids, release of the proton is enhanced by a greater number of electronegative oxygen atoms bound to the central chlorine atom. The more oxygen atoms in the structure, the greater is the stability of the conjugate anion. Recall the material of Section 7-12.

20-7B

(a) Like H_2O_2 (Figure 18-1), O_2F_2 is bent, and two independent planes defined by separate sets of the atoms OOF lie at 87.5° to one another. See Structure 18-4 in the text. This angle is typically 98° in H_2O_2 or in peroxides, although some variation is found from one compound to another. Each oxygen atom in O_2F_2 is of the AB_2E_2 classification, and the Lewis diagram is the following:

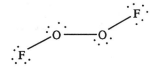

(b) The chlorine atom in ClO_2 is of the type $AB_2E_{1.5}$, and the angle O-Cl-O (117.4°) is less than the trigonal 120°, but more than the tetrahedral 109.5°. The dimer, Cl_2O_4, is shown in Structure 20-V. The Lewis diagram for ClO_2 is the following:

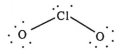

(c) Bromine in BrO_3^- is of the type AB_3E. The pyramidal geometry and the O-Br-O angle (104°) are consistent with the repulsive nature of the lone pair on the bromine atom:

$$\left[\ddot{\underset{\cdot\cdot}{O}} - Br_{/\!/\!/}\, \underset{\cdot\cdot}{\ddot{O}} \atop \underset{\cdot\cdot}{\ddot{O}} \right]^-$$

(d) This anion should more properly be written $[IO_2(OH)_4]^-$, because the Lewis diagram is the following:

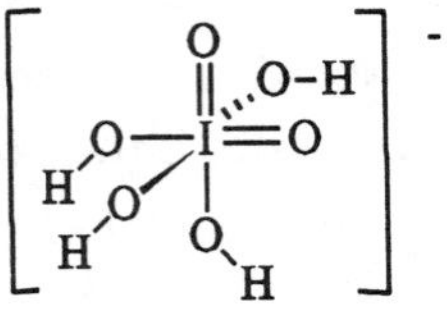

The iodine atom is of the AB_6 classification, and the I=O distances (*ca.* 1.8 Å) are typically shorter than the I-OH distances (*ca.* 1.9 Å).

(e) The ion BrO_4^- has the tetrahedral structure one would expect for a bromine atom of the type AB_4. The Br-O distance (1.61 Å) is somewhat shorter than the sum of the single bond covalent radii, which are listed in Figure 2-15.

20-9B
This can be done photochemically by taking advantage of reaction 20-7.18, and noting that NO is also a radical:

$$F_3C \cdot + \cdot NO \rightarrow F_3C-NO$$

20-11B
First consider the cation, $[I_3]^+$. The Lewis diagram is as follows:

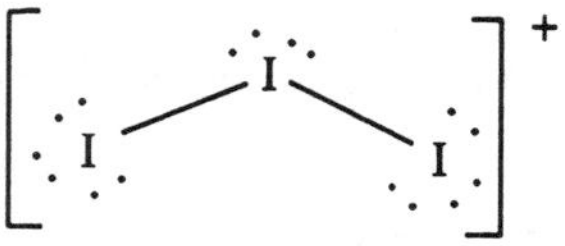

The central iodine atom is of the type AB_2E_2, and a straightforward σ bond system involving sp^3 hybridization of the central iodine atom can be invoked. The system is saturated, and the bonds are single bonds.

The Lewis diagram for the anion, $[I_3]^-$, shows the central iodine atom to fall into the classification AB_2E_3:

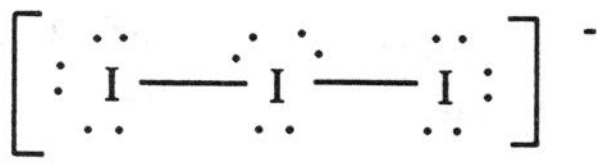

$$\left[:\overset{..}{\underset{..}{I}}—\overset{.\,.}{\underset{..}{I}}—\overset{..}{\underset{..}{I}}: \right]^-$$

The I-I bond lengths are somewhat longer than the typical I-I single bond, indicating that regular σ bonds employing a dsp^3 hybrid set on the central iodine atom may not be a realistic approach. If the central iodine atom is taken to be sp^2-hybridized, the three lone pairs can occupy the sp^2 hybrid set and achieve the maximum $120°$ angle between themselves. This allows the remaining, unhybridized p-orbital of the central iodine atom to be invested in a three-center, four-electron bond system, which is diagrammed as follows:

σ_b =

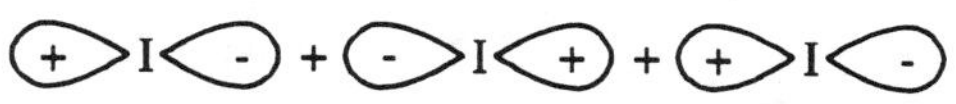

σ_n =

σ_a =

Three-center, four-electron bonds should be weaker than two-center, two-electron bonds.

CHAPTER 20

C - Problems from the Literature of Inorganic Chemistry

20-1C

(a) The electronic spectra (Figures 1 and 2 in the article) of $KBrO_4$ and $HBrO_4$ in water are the same, as are the IR and Raman spectra of the two compounds (Figures 3 and 4).

(b) Periodates are rapidly hydrated as in Figure 20-1 of the text, to give $[IO_2(OH)_4]^-$. This is also shown in the equation of the paper by Appelman. Perbromate ion does not exchange ^{18}O with enriched water, as demonstrated by mass spectrometry. This means that the equilibria (and, hence, reversible exchange of ^{18}O from water) such as those of Figure 20-1 of the text do not operate for perbromate ion.

20-3C

(a) This is best viewed as a 1:2 adduct between one BrF_5 and two SbF_5 molecules to give the ions BrF_4^+ and $Sb_2F_{11}^-$, although the two ions in the product are weakly coupled by Sb-F···Br bridges. Figure 1 depicts one repeat unit in a chain structure involving these "not-so-discrete" ions.

BrF_4^+ is a Lewis acid, being coordinated by F(5) from Sb(2) and by F(4) from an adjacent Sb(1) that is not shown.

The central Sb(2) atom may be viewed as the AB_6-type ion $[SbF_6]^-$, one fluorine of which (F(5)) forms a dative bond to Br. Another fluorine of this central SbF_6^- ion, namely F(12), forms a dative bond to the Sb(1) atom.

The SbF_5 fragment is both a Lewis acid and a Lewis Base. It forms a donor bond (through F(4)) to an adjacent Br (not shown). It also accepts a lone pair from atom F(12).

In the above analysis, it is consistently assumed that the dative bonds are somewhat longer than, and hence readily distinguished from, the typical single X-F covalent bond.

(b) In the solid state structure of $BrF_5 \cdot 2SbF_5$, which is shown in this paper to be $[BrF_4][Sb_2F_{11}]$, each Sb atom achieves an occupancy of six through the use of nearly linear fluorine bridges. Bromine achieves an effective occupancy of 7. In the resulting chain structure, the coordination geometry around the Br atom and each Sb atom is roughly octahedral, although there are differences between covalent and dative Sb-F or Br-F distances. Also it must be remembered that there is a lone pair of electrons on each Br atom. Otherwise we might expect the following geometries.

$[BrF_4]^+$, AB_4E

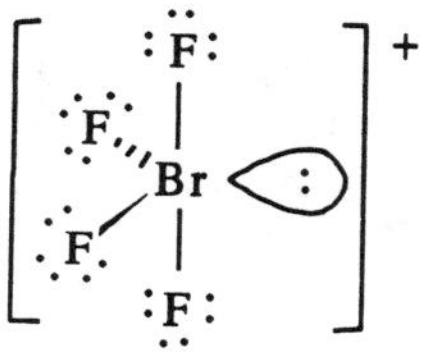

$[SbF_6]^-$, AB_6

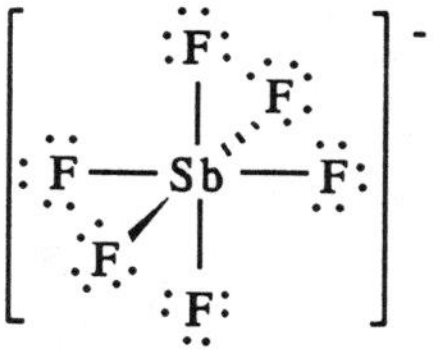

SbF_5, AB_5

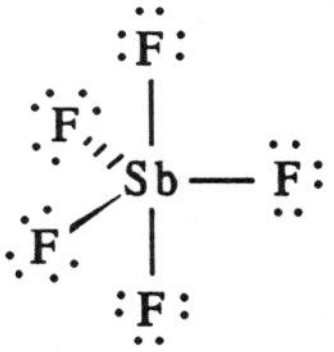

SeF_4, AB_4E

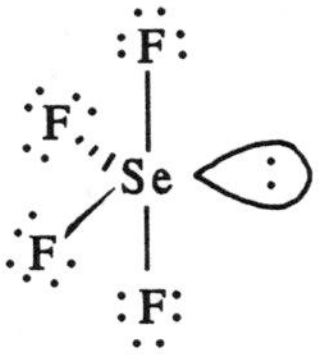

20-5C

The bond lengths that are actually observed in numerous covalent fluorides (Table I of the article) are significantly shorter than those calculated from either Pauling's radius (64 pm) or half the bond distance in difluorine (72 pm). Besides the value 0.54 pm, the authors also give fluorine atom radii to be used in three other structural situations: (a) long bonds arising from complete valence shells with lone pairs, (b) short bonds arising from incomplete valence shells without lone pairs, and (c) bonds in incomplete valence shells with lone pairs.

A - Review

21-1A

Primordial helium and the helium that is produced in the stars arise from the fusion of hydrogen. All such helium has long ago escaped the earth's gravitational force, leaving geological deposits as the only current terrestrial source. Helium arises from α particles emitted by isotopes of uranium and thorium. It is therefore found in those radioactive mineral deposits where the geological formation is able to trap the gas once it is formed.

21-3A

These three known fluorides of xenon can be prepared from the elements (or from one another), above 250 °C, according to reactions 21-2.1 through 21-2.3. If XeF_2 is desired, a deficiency of F_2 is employed. XeF_4 is obtained cleanly at 400 °C using a 1:5 mixture of Xe and F_2. Pressures above 50 atm and temperatures above 250 °C favor formation of XeF_6, either from Xe and F_2, or from XeF_4 and F_2.

21-5A

Xenates ($[HXeO_4]^-$, containing Xe^{VI}) and perxenates ($[XeO_6]^{4-}$, containing Xe^{VIII}) are formed by the action of aqueous hydroxide on XeO_3 and $[HXeO_4]^-$, respectively. The reactions are 21-2.12 and 21-2.13. Perxenate may also be obtained from oxidation of $[HXeO_4]^-$ by ozone. Finally, in water solution, acids promote the reduction of perxenate to xenate, according to equation 21-2.14.

B - Additional Exercises

21-1B

(a) $2[HXeO_4]^- + 2O_3 + 4OH^- \rightarrow 2[HXeO_6]^{3-} + 2H_2O + 2O_2$

(b) $XeO_3 + I^- \rightarrow IO_3^- + Xe$

$XeO_3 + 9I^- + 6H^+ \rightarrow Xe + 3H_2O + 3I_3^-$

(c) $XeF_2 + 2HCl \rightarrow Cl_2 + Xe + 2HF$

(d) $XeF_2 + 2Ce^{3+} \rightarrow 2Ce^{4+} + Xe + 2F^-$

(e) $XeF_2 + SbF_5 \rightarrow [XeF][SbF_6]$

   Similarly, with XeF_4, we have:

   $XeF_4 + 2SbF_5 \rightarrow [XeF_3][Sb_2F_{11}]$

21-3B

Most of the geometries are listed in Table 21-2. Lone pairs on terminal atoms (ABE_3) have been omitted for clarity.

(a) XeF_4 (AB_4E_2)

(b) XeO_3 (AB_3E_1)

(c) $[XeO_6]^{4-}$ (AB_6)

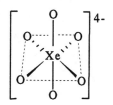

(d) $XeOF_4$ (AB_5E)

(e) $[XeF_5]^+$ (AB$_5$E)

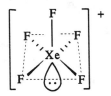

(f) $[HXeO_4]^-$ (AB$_4$E)

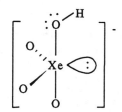

(g) $[XeF_8]^{2-}$ The geometry is that of a square antiprism, shown in Figure 6-1a of the text.

(h) $[XeF_7]^-$ The geometry is unknown, but a pentagonal bipyramid is reasonable.

21-5B

These reactions may be regarded as acid-base reactions of the Lux-Flood type, namely the transfer of oxide from the donor (base) to the acceptor (acid). An order of Lux-Flood acidities of various xenon fluorides is given in Section 21-2, following equation 21-2.5. See also the article associated with question 21-1C.

(a) $XeF_6 + XeO_2F_2 \rightarrow 2XeOF_4$

In this reaction, we consider that the acid XeF_6 accepts oxide from the base XeO_2F_2. Two fluoride ions are also transferred.

(b) $XeO_3F_2 + XeO_2F_2 \rightarrow XeOF_4 + XeO_4$

In this reaction, XeO_3F_2 serves as the acid, accepting oxide to become XeO_4. Concurrently, the oxide donor (base, XeO_2F_2) accepts two fluoride ions to become $XeOF_4$.

(c) XeF_2 is the weakest oxide acceptor in the series presented in the text. It does not accept oxide from even $[XeO_6]^{4-}$.

C - Questions from the Literature of Inorganic Chemistry

21-1C
(a) Huston has proposed an hydrolysis sequence somewhat different from that of Appleman and Malm, involving the intermediacy of $XeOF_2$:

$$XeF_4 + H_2O \rightarrow XeOF_2 + 2HF$$
$$XeOF_2 \rightarrow XeF_2 + 1/2O_2$$
$$XeOF_2 + H_2O \rightarrow XeO_2 + 2HF$$
$$XeO_2 \rightarrow XeO + 1/2O_2$$
$$XeO \rightarrow Xe + 1/2O_2$$

It is also speculated that XeO_3 is formed as follows:

$$XeOF_2 + XeO_2 \rightarrow XeO_2F_2 + XeO$$
$$XeO_2F_2 + H_2O \rightarrow XeO_3 + 2HF$$

(b) See Table 1 of the article for the list of reactions.

Reaction No.	Lux-Flood Acid (oxide acceptor)	Lux-Flood Base (oxide donor)
1	$XeOF_4$	XeO_3
2	XeF_6	XeO_3
4	XeF_6	XeO_2F_2
11	$XeOF_4$	$[XeO_6]^{4-}$
12	XeO_2F_2	$[XeO_6]^{4-}$
16	XeO_3F_2	XeO_2F_2
17	XeO_3F_2	XeO_3

146

(c) The base strength (oxide donor ability) follows the order $[XeO_6]^{4-}$ (which will transfer oxide to $XeOF_4$ and XeO_2F_2, the weakest acids) > XeO_3 (which will transfer oxide to $XeOF_4$, XeF_6, and XeO_3F_2) > XeO_2F_2.

A - Review

22-1A

This is given in Figure 2-12. Although these three elements fall at the end of the d-block of the periodic table, they are not properly regarded as transition elements, as discussed in Chapter 23. As do the elements of Group IIA(2), the elements of Group IIB(12) typically form 2+ cations. The latter also form univalent cations. The chemistry of the ions of the Group IIB(12) elements is different from that of the Group IIA(2) elements because the filled d-orbital configurations of Zn^{2+} ($[Ar]3d^{10}$), Cd^{2+} ($[Kr]4d^{10}$) and Hg^{2+} ($[Xe]5d^{10}$) make these ions very polarizable.

22-3A

(a) $Zn(s) + 2HCl(aq) \rightarrow Zn^{2+}(aq) + 2Cl^- + H_2$

(b) $Zn(s) + 2OH^-(aq) \rightarrow [ZnO_2]^{2-}(aq) + H_2$

$[ZnO_2]^{2-} + 2H_2O \rightarrow [Zn(OH)_4]^{2-}$

The net reaction is, then:

$Zn + 2OH^- + 2H_2O \rightarrow [Zn(OH)_4]^{2-} + H_2$

22-5A

The univalent ions contain metal-metal bonds, $^+M\text{-}M^+$. The electron configurations beyond the noble gas cores should, therefore, be written: $[\sigma_{ns}]^2$.

22-7A

The "hydroxide" of Hg^{I}, formed between Hg_2^{2+} and OH^- in aqueous solution, readily disproportionates to HgO and Hg. Similar behavior occurs for the sulfide. The appropriate reactions are 22-3.6 and 22-3.7.

22-9A

Only the fluoride of Hg^{2+} is an ionic substance; it has the fluorite structure shown in Figure 4-1. The chloride of Hg^{2+} has a highly covalent nature. Like the halides HgI_2 and $HgBr_2$, $HgCl_2$ is composed of linear Cl-Hg-Cl molecules. The molecules are associated in the crystalline state through chlorine bridges ($Cl_2Hg\cdots Cl\text{-}Hg\text{-}Cl$) between

adjacent molecules. This bridging gives each Hg^{II} a nearly
octahedral environment composed of two axial chlorine atoms
(covalently bound) and four equatorial chlorine atoms, which are
bridging atoms.

B - Additional Exercises

22-1B

For the reaction:

$$Hg(l) + 1/2O_2(g) \rightarrow HgO(s)$$

ΔG is negative below 400 °C. The enthalpy for the reaction is
negative, *i.e.* the process is exothermic. The entropy of reaction
is, however, negative (unfavorable), because one mole of a liquid
plus 1/2 mole of a gas condense to a solid. Also, $Hg(l)$ boils at
358 °C. These factors cause the term $T\Delta S$, which is negative, to
predominate at high temperatures, causing ΔG (which is equal to
the quantity $\Delta H - T\Delta S$) to become positive. See also the answer to
question 1-5(a), Chapter 1.

22-3B

First, Hg_2^{2+} should be diamagnetic because of its Hg-Hg single bond,
as described in the answer to question 22-5A. Second, the
symmetrical, linear ions show a stretching vibration in the Raman
spectra. (The symmetrical vibration is not infrared active.) The
conductivities of solutions of Hg^I halides are more in keeping with a
formulation Hg_2X_2 than with the formulation HgX. Measurements of
colligative properties of aqueous solutions of Hg^I halides are also
in agreement with the formulation Hg_2X_2, not HgX. Hg_2X_2 salts ionize
upon dissolution according to the equation:

$$Hg_2X_2(s) \rightarrow Hg_2^{2+}(aq) + 2X^-(aq)$$

22-5B

(a) $Hg_2^{2+} + 2OH^- \rightarrow Hg + HgO + H_2O$

(b) $HgF_2 + H_2O \rightarrow HgO + 2HF$

(c) $Zn^{2+} + H_2O \rightarrow Zn(OH)^+ + H^+$

(d) $HgO \rightarrow Hg + 1/2O_2$

C - Questions from the Literature of Inorganic Chemistry

22-1C

As made clear in this review, we must distinguish between two types of bonds between mercuric ion and thiolate ligands. These are (1) primary bonds to terminal thiolate ligands and (2) secondary bonds to bridging thiolate ligands. The bonds to bridging thiolate ligands (secondary bonds) are longer than the primary bonds. The overall result is that there are two types of coordination in the solid state structures, as set down in Section B of the article. The authors further list the structural details for both mononuclear and polynuclear complexes with coordination number two (Table III), coordination number three (Table IV) and coordination number four (Table IV). See also Figures 3-8 of the article. The general conclusion is that mercury(II) thiolates often adopt structures with coordination numbers in the solid state that are larger than might otherwise be guessed from their formulas. This is accomplished by the ability of a thiolate ligand to serve as a ligand bridge in much the same way as the halogens are known to do among the halides of the elements.

A - Review

23-1A

The transition elements have been defined in Section 23-1 to be those elements that have a partially filled d or f shell, either for the element itself or for any of its chemically important oxidation states. This definition includes Cu, Ag, and Au of Group IB(11), but excludes the metals of Group IIB(12), Zn, Cd, and Hg. This means that at least 50% of the elements are transition elements.

23-3A

The orbital shapes are given in Figure 23-2. In an octahedral ligand field, the e_g set is is composed of the d_{z^2} and $d_{x^2-y^2}$ orbitals. The t_{2g} set is composed of the orbitals d_{xy}, d_{yz}, and d_{zx}.

23-5A

This is shown in Figure 23-6, where an octahedral ligand field diagram becomes a ligand field diagram appropriate for a four-coordinate, square complex.

23-7A

As shown in Figures 23-13 and 23-14, only the configurations d^4, d^5, d^6, and d^7 give both a low spin and a high spin case in an octahedral ligand field.

23-9A

There is one major selection rule that outright forbids d-d electronic transitions in octahedral complexes. This is the selection rule that d-d transitions are parity forbidden. It is a selection rule that does not apply to tetrahedral complexes, or to other geometries lacking a center of symmetry. This happens because all d orbitals have *gerade* symmetry with respect to inversion, and consequently, no d-d transition can accomplish the change in parity (g → u or u → g) that must occur to make the transition allowed. Additionally, d-d transitions may be spin-forbidden as well as orbital-symmetry forbidden.

23-11A

The spectrochemical series is the relationship between the size of the ligand field splitting parameter Δ and the identity of the ligand that causes the splitting. The various ligands are arranged in the order of their increasing ability to cause the d-orbital splittings, as listed in Section 23-7. The order is not always rigorously obeyed; for instance, it may change depending on the oxidation state of the metal.

23-13A

(a) The value of Δ_o depends on the oxidation state of the central metal, the higher oxidation state corresponding to the larger value of Δ_o. This is mostly due to the more severe electrostatic interaction of the ligand set with a higher metal positive charge.

(b) The value of Δ_o increases about 30-50% upon replacing a first-series metal ion with a corresponding second-series metal ion, provided the two have the same oxidation state. One important consequence is that the complexes of second- and third-row transition metal compounds are more frequently low spin than high spin.

23-15A

First notice the open circle used to plot the value of the Cr^{2+} ionic radius in Figure 23-24. Because of Jahn-Teller distortions, Cr^{2+} complexes do not have perfectly octahedral coordination geometries. The Cr^{2+} ion is a d^4 system, specifically $(t_{2g})^4$. Because the t_{2g} set of orbitals would be degenerate in a perfectly octahedral ligand environment, and as required by the Jahn-Teller effect, Cr^{2+} coordination compounds are tetragonally distorted.

B - Additional Exercises

23-1B

The axes defined in Figure 23-5 will also be used here. The two changes in the tetrahedral crystal field diagram are the inverse of one another.

(a) For the tetrahedron that is flattened on the z axis:

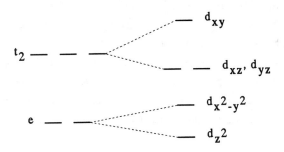

(b) For a tetrahedron that is elongated on the z axis:

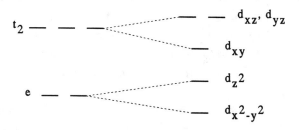

23-3B

A cube (eight vertices) and one tetrahedron (four vertices) are superimposed in Figure 23-5. From this it can be seen that two tetrahedra may be superimposed so that their vertices define a cube. The orbital splitting for the cube is, therefore, simply twice that of a tetrahedron:

$$2\Delta_t = \Delta_{cube}$$

23-5B

As shown in Figure 23-17, the value of Δ_o for $[Ti(H_2O)_6]^{3+}$ is ca. 20,000 cm^{-1}. Compared to water, Cl^- is lower and CN^- is higher in the spectrochemical series. We expect that the value of Δ_o for the complex $[TiCl_6]^{3-}$ should be smaller than that for $[Ti(H_2O)_6]^{3+}$. Also, in a simplistic analysis, we expect that the value of Δ_o for the complex $[Ti(CN)_6]^{3-}$ should be larger than that for the hexaaquo complex.

This simplistic approach has limitations. For instance, the spectrum shown in Figure 23-17 clearly has a shoulder on the low energy side of the absorption maximum, most likely due to Jahn-Teller distortions. There will certainly be differences in bonding to Ti^{III} among such a wide variety of ligands, and there may even be differences in coordination number. For example, the cyanide complex that predominates in solution may be $[Ti(CN)_8]^{5-}$.

23-7B
Table 23-1 is pertinent to the discussion.

(a) Although Fe^{3+} is a d^5 ion, cyanide is a strong field ligand, and we expect that Δ_0 is very much larger than 30,000 cm^{-1}. (The value of Δ_0 for $[Fe(CN)_6]^{4-}$ is 33,000 cm^{-1}.) A low spin complex is predicted. The LFSE for low spin d^5 cases is 5 x $2/5\Delta_0$ = $2\Delta_0$. There is one unpaired electron.

(b) This is a d^6 system. Ruthenium is a second row transition element, which should make Δ_0 large. We predict a low spin case. The ligand field stabilization energy for low spin d^6 systems is 6 x $2/5\Delta_0$, and the complex is diamagnetic.

(c) As shown in Table 23-1, $[Co(NH_3)_6]^{3+}$ is a low spin case because Δ_0 (23,000 cm^{-1}) is greater than the pairing energy P. The LFSE is 6 x $2/5\Delta_0$. The electrons are all paired, and the ion is diamagnetic.

(d) The tetrachlorocobaltate(II) ion is a d^7 system. As shown in Table 23-1 for Co^{II}, P > Δ_0, even for the ligand H_2O. The chloride ligand is lower in the spectrochemical series than is water, and the tetrachlorocobaltate(II) complex is high spin. There are three unpaired electrons, and the LFSE is $6/5\Delta_t$.

(e) As shown in Table 23-1, for $[Fe(H_2O)_6]^{2+}$, P (17,600 cm^{-1}) > Δ_0 (10,400 cm^{-1}), and the ion is high spin. The LFSE is $2/5\Delta_0$, and there are four unpaired electrons.

(f) This high spin case is a d^5 system, for which the LFSE is zero. There are five unpaired electrons.

(g) The d^6 ion $[CoF_6]^{3-}$ is high spin, and the LFSE is $2/5\Delta_o$. There are four unpaired electrons.

(h) The d^4 ion $[Cr(H_2O)_6]^{2+}$ is high spin, and the LFSE is $3/5\Delta_o$. There are four unpaired electrons.

23-9B

The donor orbital is a filled t_{2g} metal orbital. The acceptor orbital is the empty π^* orbital of CO, which is π_2 in Figure 3-27b.

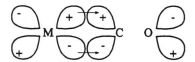

23-11B

The d^8 Tanabe-Sugano diagram is the proper one here, and the actual spectrum is given in Figure 23-22. We choose a value of $\Delta/B = 13$, because, as described in the text, this value on the x-axis of the Tanabe-Sugano diagram gives a proper correspondence between the ratio of the wave numbers for any two of the three absorption bands. The value of B for the free ion is 1080 cm^{-1}, and 80% of that (865 cm^{-1}) is a reasonable estimate of the value of B in the tris(ethylenediamine) complex. Using the point on the diagram where $\Delta B = 13$, we can judge the value of E/B for each spin-allowed excited state:

State	Energy (E/B)	Energy (cm^{-1})
$^3T_2$	14	1.2×10^4
$^3T_1$	21	1.8×10^4
$^3T_1$	34	2.9×10^4

These values should be compared with those available from Figure 23-22.

A - REVIEW

24-1A

Ti^{IV}, $[Ar]4s^0 3d^0$

V^{II}, $[Ar]3d^3$

Cr^V, $[Ar]3d^1$

Mn^{VI}, $[Ar]3d^1$

Fe^0, $[Ar]4s^2 3d^6$ or $[Ar]3d^8$

Co^I, $[Ar]4s^1 3d^7$ or $[Ar]3d^8$

Ni^{II}, $[Ar]3d^8$

Cu^{III}, $[Ar]3d^8$

Ti^{III}, $[Ar]3d^1$

24-3A

Jahn-Teller distortions (of octahedral geometries) stemming from the t_{2g} orbital set are possible for the following: d^1, d^2, low spin d^4, low spin d^5, high spin d^6, and high spin d^7. The effect is small because the t_{2g} electrons are not oriented towards the ligands. Jahn-Teller distortions (of octahedral geometries) stemming from the e_g orbital set are possible for the following: high spin d^4, low spin d^7, and d^9. These distortions are more pronounced than the ones mentioned above because the e_g orbitals are pointed directly towards the ligands, and a tetragonal distortion greatly affects the energy of the e_g orbitals. From the list in question 24-2A, we predict Jahn-Teller distortions for Cr^V, Mn^{VI}, and Ti^{III}, all of which are d^1 ions. The effects are usually small, though, for the reasons noted above. The largest Jahn-Teller distortions typically occur for complexes of Cu^{II}, a d^9 ion.

24-5A

The compounds MCl_4 serve as Lewis acids in forming the compounds $[TiCl_4(OPCl_3)]_2$, $TiCl_4(OPCl_3)_2$, and $[TiCl_6]^{2-}$. The compounds MCl_3 serve as Lewis acids in forming the compounds $VCl_3(NMe_3)_2$, $CrCl_3 \cdot (THF)_3$, $[VCl_4]^-$, $[CrCl_4]^-$, and $[FeCl_4]^-$.

24-7A

The pentacoordinate compound $VO(acac)_2$, Structure 24-II, adds a sixth ligand readily. The compound formed in pyridine solvent is $VO(acac)_2py$. The presence of the sixth ligand influences the extent of $O \rightarrow V$ π donation, and this changes the V-O bond strength and the V-O stretching frequency.

24-9A

The spinels have the general formula AB_2O_4, and examples are $MgAl_2O_4$ (the prototype), $Mn^{II}(Mn^{III})_2O_4$, and Mg_2TiO_4. The alums are sulfates such as $KAl(SO_4)_2(H_2O)_{12}$ (the prototype), $CsTi(SO_4)_2 \cdot 12H_2O$ and $KV(SO_4)_2 \cdot 12H_2O$. The perovskites are mixed metal oxides such as $CaTiO_3$ (the prototype), $SrTiO_3$, and $BaTiO_3$. Salts of the anion MnX_3^- also have the perovskite structure.

24-11A

The dimer $Cr_2(CO_2Me)_4(H_2O)_2$ has a structure analogous to the copper derivative, structure 24-X. Each bridging acetate group may be represented as:

Atom	AB_xE_y Classification	Hybridization
Water oxygen atom	AB_3E	sp^3
Acetate C-O oxygen atom	AB_2E_2	sp^3
Acetate C=O oxygen atom	AB_2E	sp^2
Acetate CO_2 carbon atom	AB_3	sp^2
Acetate CH_3 carbon atom	AB_4	sp^3

The trimer cation has the structure shown in 24-I. The water and acetate ligands are completely analogous to those in the dimer above. The central oxygen atom has a nearly planar arrangement of Cr atoms around it, and together they constitute an AB_3 system. In this respect, the central oxygen atom is like that of H_3O^+.

24-13A

(a) $Mn(OH)_2$ is oxidized by atmospheric oxygen to Mn_2O_3 and eventually to MnO_2. See equation 24-22.1.

(b) There are two modifications of cobalt(II) hydroxide. The pink form is precipitated from aqueous Co^{II} solutions upon addition of alkali metal hydroxides. It darkens in air. The blue form of cobalt(II) hydroxide is obtained from aqueous solutions by employing excess Co^{II} salts.

(c) The blue hydroxide of Cu^{II} is readily dehydrated to the black oxide:

$$Cu(OH)_2 \rightarrow CuO + H_2O$$

24-15A

(a) Spin-paired Mn^{2+}: $[Mn(CN)_6]^{4-}$.
(b) Tetrahedral Cr^{4+}: $Cr(OBu-t)_4$ and $Cr(CH_2SiMe_3)_4$.
(c) Tetrahedral Co^{2+}: $[Co(H_2O)_4]^{2+}$ and $[CoCl_4]^{2-}$.
(d) Octahedral V^{3+}: $[V(ox)_3]^{3-}$.
(e) Octahedral Co^{3+}: $[Co(NH_3)_6]^{3+}$.
(f) High-spin Fe^{2+}: $[Fe(H_2O)_6]^{2+}$.
    Low-spin Fe^{2+}: $[Fe(CN)_6]^{4-}$.
(g) High-spin Mn^{2+}: $[Mn(H_2O)_6]^{2+}$ and $[MnCl_6]^{4-}$.

24-17A

We must consider two geometrical isomers (*cis* and *trans*) and two possibilities for linkage isomerism (N-bound and S-bound) of each thiocyanate ligand. Let us use the nitrate salt for the following answers.

cis-[Co(en)$_2$(-SCN)$_2$]NO$_3$:
cis-dithiocyanatobis(ethylenediamine)cobalt(III) nitrate

cis-[Co(en)$_2$(-NCS)$_2$]NO$_3$:
cis-diisothiocyanatobis(ethylenediamine)cobalt(III) nitrate

cis-[Co(en)$_2$(-NCS)(-SCN)]NO$_3$:
cis-isothiocyanatothiocyanatobis(ethylenediamine)cobalt(III) nitrate

There are also three similarly named *trans* isomers.

B - Additional Exercises

24-1B
(a) From equation 24-7.4, we have the following geometries.
Ti(NEt$_2$)$_4$ has a tetrahedral titanium atom and trigonal nitrogen atoms.

$$Et_2N \underset{Et_2N}{\overset{NEt_2}{\underset{}{\diagdown}}} Ti \underset{NEt_2}{\overset{}{\diagup}}$$

Carbon disulfide is linear.

$$\ddot{S} = C = \ddot{S}$$

Each dithiocarbamate in Ti(S$_2$CNEt$_2$)$_4$ is bidentate, and titanium has a dodecahedral coordination geometry. The geometry of each anionic dithiocarbamate ligand is given in Structure 14-IV.

(b) From equation 24-14.2, we have the following geometries. Dichromate is depicted in Figure 24-2. Carbonate is triangular, and CO is obviously linear. The geometry of carbon either as graphite or as diamond is discussed in Chapter 8. Chromium(III) oxide, Cr$_2$O$_3$, in its α form, has the corundum (α-Al$_2$O$_3$) structure as discussed in Chapter 4.

(c) For equation 24-18.1, we have the following geometries. $CrCl_3 \cdot (THF)_3$ is octahedral, and THF is the cyclic ether OC_4H_8. The anion $[CH_2SiMe_3]^-$ may be regarded as a Lewis base,

$$\left[\: :CH_2 - SiMe_3 \right]^-$$

in which each carbon is pyramidal and Si is tetrahedral. the coordination geometry in $[Cr(CH_2SiMe_3)_4]^-$ is likely tetrahedral.

(d) For equation 24-25.7, we have the following geometries. $[MnO_4]^-$ is tetrahedral whereas $[MnO_3]^+$ is trigonal planar. Sulfuric acid has tetrahedral sulfur, as does bisulfate:

The geometry of H_3O^+ is depicted in Structure 7-1.

(e) For reaction 24-33.7 we may have either *cis* or *trans* coordination geometry for the cobalt(III) complexes:

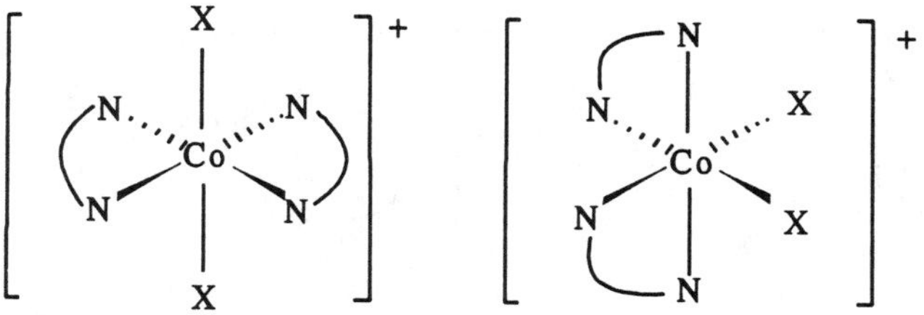

The Lewis diagram for the linear thiocyanate ion is:

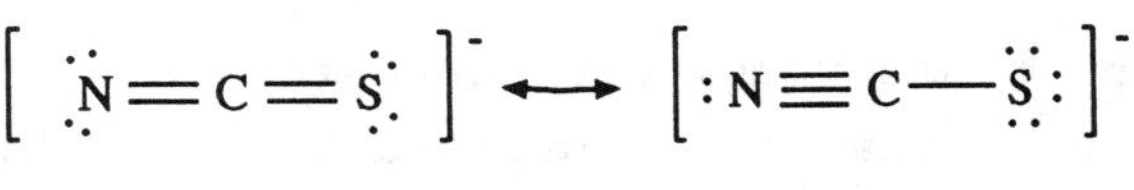

24-3B

The densities of the metals of the first transition series increase somewhat regularly from left to right across the row. The trend is roughly the inverse of the variation in the atomic volumes. Because density is a property of a bulk sample, and not of an individual atom, the trend should not be interpreted with as much detail as that, say, of ionization enthalpy or ionic radius (Chapter 23). Nevertheless, it is apparent that the metal densities increase both because the metallic bond becomes stronger (more electrons) and because the atomic size decreases due to imperfect shielding. See also the discussions on metal bonding in Chapters 8 and 32.

24-5B

This tris-chelate complex undergoes a tetragonal Jahn-Teller distortion.

24-7B

In particular, for K_2CuCl_3, the structure in the solid has chains built up through Cl atom bridging. In general, especially for halides, the solid structures often have three dimensional networks built up through bridging halogen atoms. One must pay particular attention to the actual solid state structure, remembering that stoichiometry may not be predictably related to structure.

24-9B

C - Questions from the Literature of Inorganic Chemistry

24-1C

(a) Two independent geometries were found for the $[Ni(CN)_5]^{3-}$ ion in this salt: square pyramidal and distorted trigonal bipyramidal.

(b) For the square pyramid, the hybridization of Ni^{II} (Figure 3-6) is dsp^3, and the crystal field diagram is:

$$— \quad d_{xy}$$

$$\underline{\downarrow\uparrow} \quad d_{z^2}$$

$$\underline{\downarrow\uparrow} \quad d_{x^2-y^2}$$

$$\underline{\downarrow\uparrow} \; \underline{\downarrow\uparrow} \quad d_{xz}, d_{yz}$$

For the distorted trigonal bipyramid, the hybridization should be qualitatively the same as that of a regular trigonal bipyramid (dsp^3, Figure 3-6). The crystal field diagram is:

$$— \quad d_{z^2}$$

$$\underline{\downarrow\uparrow} \; \underline{\downarrow\uparrow} \quad d_{xy}, d_{x^2-y^2}$$

$$\underline{\downarrow\uparrow} \; \underline{\downarrow\uparrow} \quad d_{xz}, d_{yz}$$

Note that the distortion observed in this paper from a regular trigonal bipyramid involves one long equatorial Ni-CN length, and an enlarged equatorial bond angle ($141°$). For more discussion, see the article of question 24-3C in the text.

(c) The square pyramidal and the distorted trigonal bipyramidal geometries were found in the same crystal, and, hence, must not be significantly different in stability.

(d) Compare Figures 3 and 2 in the article. In order to convert the structure of Figure 3 into that of Figure 2, it is necessary only to move two CN^- ligands (numbers 2 and 4 in Figure 3) and to lengthen the bond to one CN^- ligand (number 1 in Figure 3).

(e) Three processes are required.

1. oxidation of Co^{II}:

$$4Co^{2+} + 4H^{+} + 12en + O_2 \rightarrow 4[Co(en)_3]^{3+} + 2H_2O$$

2. addition of cyanide to tetracyanonickelate(II):

$$CN^{-} + [Ni(CN)_4]^{2-} \rightarrow [Ni(CN)_5]^{3-}$$

3. co-precipitation of the mixed salt:

$$[Co(en)_3]^{3+}(aq) + [Ni(CN)_5]^{3-}(aq) \rightarrow [Co(en)_3][Ni(CN)_5](s)$$

24-3C

(a) These authors conclusively demonstrated that the ion $[Cu^{II}Cl_5]^{3-}$, as the hexaamminechromium(III) salt, has trigonal bipyramidal geometry with axial Cu-Cl bond lengths that are significantly shorter than the equatorial ones.

(b) The crystal field diagrams are:

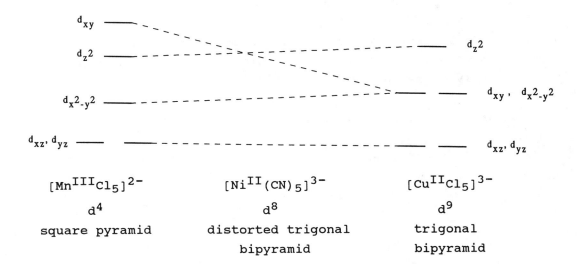

$[Mn^{III}Cl_5]^{2-}$	$[Ni^{II}(CN)_5]^{3-}$	$[Cu^{II}Cl_5]^{3-}$
d^4	d^8	d^9
square pyramid	distorted trigonal bipyramid	trigonal bipyramid

(c) This study has confirmed Gillespie's prediction (see reference 34 in the article) for low spin d^8 or d^9 systems, that axial bond lengths in a trigonal bipyramidal complex should be shorter than equatorial ones. The prediction was based on VSEPR arguments, noting the empty (d^8) or only half-filled (d^9) d_{z^2} orbital in the trigonal bipyramidal crystal field diagram shown in the answer to part b above. The distortion from a square pyramid (e.g. $[MnCl_5]^{2-}$) to a trigonal bipyramid (e.g. $[CuCl_5]^{3-}$) switches the positions of the d_{z^2} and the d_{xy} orbitals.

24-5C

(a) Since the trimethylamine oxidation products and the identity of the reducing agent have not been determined, we may only write a half-reaction:

$$VOCl_3 + 2NMe_3 + e^- \rightarrow VOCl_2 \cdot 2NMe_3 + Cl^-$$

(b) The magnetic susceptibility data ($\mu_{eff} = 1.74$ BM) indicate one unpaired electron (Table 2-4 of the text), as expected for V^{IV}.

(c) This is essentially the diagram of Figure 3-14 in the text. The d_{xy} orbital of V^{IV} accepts electron density from oxygen. This constitutes $(3d_{xy})\pi \rightarrow (2p_x)\pi$ overlap.

(d) The V-O distance (1.59 ± 0.02 Å) is significantly shorter than the V-O single bond (1.98 Å) found in $V^{III}(acac)_3$. It is comparable to other known V-O double bonds, e.g. that in $VOCl_2(OPPh_3)_2$ (1.58 Å).

(e) Usually one finds square pyramidal geometry for oxovanadium(IV) compounds, examples being $VOCl_2(OPCl_3)_2$ and Structure 24-II in the text. The title compound, $VOCl_2 \cdot 2NMe_3$, however, adopts the trigonal bipyramidal geometry, according to the authors of this paper, principally because of the trigonal symmetry of the trimethyl amine ligands. The authors point out that V-O multiple bonding would normally favor square pyramidal geometry.

24-7C

(a) The general transition state in an I_d process is shown in Figure 6-5c. Since the activated complex is solvated by more solvent molecules in the second coordination sphere than by entering anions X^- in the second coordination sphere, the statistical factors alone favor more frequent solvent exchange than anation. All anions should move from the second to the inner coordination sphere in the activated complex with first order rate constants that are (1) independent of the identity of the anion and (2) less than that for solvent exchange.

(b) The anation of cis-[Co(en)$_2$NO$_2$DMSO]$^{2+}$ by either Cl^- or NO_2^- is characterized by the sort of saturation kinetics (Figure 1 in the article) that are typical of an interchange mechanism, equation 4. Furthermore, these limiting rate constants are well below that for solvent exchange ($k_{DMSO} = 3.0 \times 10^{-4}$ s^{-1} at 35 °C). Also, the values of the limiting first order rate constants for Cl^- and NO_2^- substitution are close to one another, indicating a lack of dependence on the identity of the incoming ligand, as expected for a dissociative process.

(c) The rate of DMSO exchange is greater than that of X^- substitution, despite the fact that the equilibrium shown in reaction (2) favors the products, because the Co-X bond is generally stronger than the Co-DMSO bond.

A - Review

25-1A

(a) Because of the Lanthanide contraction, the radii of the third transition series elements are comparable to those of the second transition series elements, both being larger than those of the first transition series metals.

(b) Higher oxidation states of the metals of the second and third transition series are more stable than those of the first transition series metals. In some cases, the highest oxidation state of a second or third series element is the preferred oxidation state. The 2+ and the 3+ oxidation state of the second and third transition series metals are less important than for the first transition series.

(c) Metal-metal bonding is a more prominent feature of compounds of the third and second transition series metals than of the first transition series metals.

(d) Metals of the second and third transition series (excluding perhaps those of the platinum group metals) have coordination numbers that are greater than six more commonly than do the metals of the first transition series.

(e) Because the size of the crystal field splitting parameters increase for the metals of the transition series in the order 3d < 4d < 5d, second and third transition series metal compounds are more frequently low spin than are the comparable complexes of the first transition series elements.

25-3A

The metals that form six-membered octahedral clusters of the low valent ions are Nb and Ta ($[M_6X_{12}]^{n+}$, Figure 25-2), and Mo^{II} and W^{II} ($[M_6Cl_8]^{4+}$, Figure 25-4). The triangular cluster $[Re^{III}Cl_3]_3$ has a chlorine-linked polymeric structure, Figure 25-5.

25-5A

The tetraoxides of Ru and Os are volatile. OsO_4 is obtained by oxidation of the metal with O_2, whereas ruthenium reacts with oxygen to give RuO_2. Therefore, ruthenium tetraoxide requires oxidizing agents such as MnO_4^-, Cl_2, or hot $HClO_4$. OsO_4 is colorless, and RuO_4 is orange. Both molecules are tetrahedral, and are powerful oxidants. For more details, see Section 25-17.

25-7A

The Dihalides of molybdenum and tungsten have the formula M_6Cl_{12}, and the structures (Figure 25-4) contain $[M_6Cl_8]^{4+}$ clusters in which eight chlorine atoms cap the eight faces of an M_6 octahedron.

25-9A

These complexes have structures similar to that shown in Figure 25-10. The crystals are spectroscopically dichroic, and they conduct electricity most readily along that crystal axis which coincides with the M..M..M.. chains. Thus, an interaction between metal centers is indicated, even though the M-M distance is not short enough to support bonding.

25-11A

This is a bis(acetato)-bridged trimer in which each Pd^{II} center is surrounded by a planar array of four acetate atoms. (See the report by A. C. Skapski and M. L. Smart, *Chem. Commun.* **1970**, 658.)

25-13A

Although the electron configurations of Cu, Ag, and Au are analogous, and although the metals are characteristically difficult to oxidize, there are important differences in the chemistry of copper compounds as compared to those of Ag or Au. Details are given in Section 25-31.

25-15A

(a) $4Au^O + 8CN^- + O_2 + 2H_2O \rightarrow 4[Au^I(CN)_2]^- + 4OH^-$

(b) $AgI(s) + 2S_2O_3^{2-} \rightarrow [Ag(S_2O_3)_2]^{3-} + I^-$

(c) $S_2O_8^{2-} + 8py + 2Ag^+ \rightarrow 2[Ag(py)_4]^{2+} + 2SO_4^{2-}$

CHAPTER 25

25-17A

The ores are concentrated by floatation, wherein the finely powdered
sulfide ore particles are treated with an oil and a surfactant, and
are skimmed from a froth of bubbles (created by forced aeration of
the suspension). As a result, the denser rocks sink to the bottom of
the floatation chamber. After refining the ore to obtain semi-pure
Ni-Cu matte, the alloy is molded into anodic rods. Electrolysis in
dilute H_2SO_4, with pure copper cathodes, disintegrates the anode, and
pure copper is deposited on the cathode. The metals in the anode
that are less electropositive than copper (namely the platinum group
metals) are not oxidized. They form a sludge in the vicinity of the
disintegrating anode. The sludge (called anodic slime) may be worked
up in a variety of ways to obtain the platinum group metals.

B - Additional Exercises

25-1B

Because the f orbitals shield one another poorly from the effects of
the growing nuclear charge on traversing the row from La to Lu, the
Lanthanides experience a dramatic increase in effective nuclear
charge with increase in atomic number. This causes a progressive
shrinking (termed the Lanthanide contraction) in the atomic and ionic
radii of the elements from La to Lu. In turn, subsequent elements
have dramatically smaller radii than might be otherwise expected
based solely on atomic number. See Figure 27-1, Table 26-1 and
Section 8-12.

25-3B

(a) $[ZrF_6]^{2-}$. This is simply octahedral.

(b) $[ZrF_7]^{3-}$. As the sodium salt, this has the geometry shown in
Structure 6-X, a pentagonal bipyramid. As the ammonium salt, it has
the geometry shown in Structure 6-XI, a face-capped octahedron.

(c) $[Zr_2F_{12}]^{4-}$. This is composed of pentagonal bipyramids sharing an
edge.

(d) $[ZrF_8]^{4-}$. This is the square antiprism, Figure 6-1A.

(e) $[Zr_2F_{14}]^{6-}$. This has the structure of square antiprisms sharing an edge.

25-5B

For the rhodium dimer $[Rh(CO)_2Cl]_2$ (Figure 25-9), these are reactions 25-25.2 through 25-25.4. The products are dicarbonyl derivatives with cis-$(CO)_2$ ligands. For platinum, there are reactions analogous to that shown in equation 25-27.1.

25-7B

In a manner analogous to that for ruthenium (problem 25-6B), Rh_2O_3 may be treated with HCl(aq). When the solution is evaporated, "$RhCl_3 \cdot nH_2O$" is obtained. Various aqua and chloro complexes are obviously present; chloride is not precipitated by Ag^+ from fresh solutions of this product. Reactions are given in Figure 25-8.

(a) Boiling with aqueous HCl gives the hexachlororhodate(III) ion, $[RhCl_6]^{3-}$.

(b) With an excess of PPh_3 in ethanol, $Rh^ICl(PPh_3)_3$ is formed. Triphenylphosphine oxide is a side product.

(c) Heating in ethanol/NH_3 gives the hexaammine, $[Rh(NH_3)_6]Cl_3$.

(d) This gives the acetato-bridged dimer of rhodium, cf. $M_2(CH_3CO_2)_4$, Structure 24-X.

25-9B

(a) $Zr^{4+} + 4OH^- \rightarrow ZrO_2 + 2H_2O$

(b) $ZrCl_4 + H_2O \rightarrow ZrOCl_2 + 2HCl$

(c) $Mo_6Cl_{12} + 6Cl_2 \rightarrow 6MoCl_4$

(d) $2Mo + 5Cl_2 \rightarrow Mo_2Cl_{10}$

(e) See Figure 25-3.

(f) $Cr_2(CO_2Me)_4 + 4HCl \rightarrow 2Cr^{2+} + 4Cl^- + 4CH_3CO_2H$

(g) $Mo_2(CO_2Me)_4 + 8HCl \rightarrow [Mo_2Cl_8]^{4-} + 4H^+ + 4CH_3CO_2H$

(h) $Mo_2Cl_{10} + 2H_2O \rightarrow 2[MoOCl_5]^{2-} + 4H^+$

(i) $MoO_3 + 4HCl \rightarrow [MoO_2Cl_4]^{2-} + H_2O + 2H^+$

(j) $Re_2O_7 + 2OH^- \rightarrow 2ReO_4^- + H_2O$

$2H^+ + 2ReO_4^- + 7H_2S \rightarrow Re_2S_7 + 8H_2O$

(k) $Re_2Cl_{10} + 2H_2O \rightarrow 2[ReOCl_5]^{2-} + 4H^+$

(l) $ReOCl_3(PPh_3)_2 + EtOH \rightarrow ReOCl_2(OEt)(PPh_3)_2 + HCl$

(m) $RuO_4 + 8HCl \rightarrow "RuCl_3 \cdot nH_2O" + 5/2Cl_2 + 4H_2O$

$\qquad\qquad\qquad\qquad\qquad$ + chloride oxidation products

(n) $2[IrCl_6]^{2-} + 2I^- \rightarrow 2[IrCl_6]^{3-} + I_2$

(o) $Rh + 3/2Cl_2 + 3NaCl \rightarrow Na_3[RhCl_6]$

$2[RhCl_6]^{3-} + 6OH^- \rightarrow Rh_2O_3(s) + 3H_2O + 12Cl^-$

$9H_2O + Rh_2O_3 + 6HClO_4 \rightarrow 2[Rh(H_2O)_6]^{3+} + 6ClO_4^-$

(p) $[Rh(CO)_2Cl]_2 + 2py \rightarrow 2cis\text{-}Rh(CO)_2Cl(py)$

(q) $PdCl_2 + 2Cl^- \rightarrow [PdCl_4]^{2-}$

25-11B

This is similar to the copper derivative, Structure 24-X, without the two terminal water ligands.

25-13B

This has a structure similar to Nb_2Cl_{10}, Figure 25-1B. It is composed of two edge-sharing octahedra.

25-15B

Compound	Oxidation State
$Mo_2(CO_2Me)_4$	Mo^{II}
$Zr(acac)_4$	Zr^{IV}
Mo_2Cl_{10}	Mo^V
$RhCl(CO)(PPh_3)_2$	Rh^I

25-17B

(a) There is $d\pi$-$p\pi$ back bonding between Ru and N_2 in $[Ru(NH_3)_5(N_2)]^{2+}$. Ruthenium (t_{2g}^6) is a good π-electron donor. The π-acceptor orbital on the N_2 ligand is the lowest unoccupied π^* orbital, essentially π_2 of Figure 3-27a.

(b) An octahedral Ru^{IV} ion whould be a low spin d^4 system, that is (t$_{2g}$)^4, with two unpaired electrons. Spin pairing of eight d electrons in a metal-metal unit (Ru$_2$) is accomplished by the formation of two perpendicular Ru(dπ)-O(pπ)-Ru(dπ) three-center, four-electron bonds, one in the xz plane and the other in the yz plane:

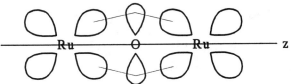

(c) Both CO and PPh$_3$ ligands in Ru(H)(Cl)(CO)(PPh$_3$)$_3$ are effective π-acceptor ligands. Ruthenium(II) (t$_{2g}$)^6 back donates electron density into the lowest unoccupied π* orbital of CO and into an empty 3d orbital of P.

C - Questions from the Literature of Inorganic Chemistry

25-1C

(a) Nb^0 + 2Nb$_2$Cl$_{10}$ → 5NbCl$_4$

(b) The structure is a dichloro-bridged chain in which there are two distinctly different Nb-Nb distances in alternating fashion: 3.029 Å and 3.794 Å. The shorter distance indicates Nb-Nb bondng between alternate metal atom pairs. The bonds to the chlorine atom bridges also alternate between 2.523 Å and 2.425 Å. As shown in Figure 1 of this article, the terminal (axial) chlorine atoms bend away from each Nb-Nb metal-metal bond.

(c) The low magnetic susceptibility indicates pairing of the two d^1 electrons in Nb-Nb sets.

(d) In NbCl$_5$, the Nb^V-Nb^V distance in the dimers Nb$_2$Cl$_{10}$ is 3.951 Å, which constitutes a non-bonded distance. This is, of course, a d^0 system, without a M-M bond.

(e) $(NbCl_5)_2$ has the structure shown in Figure 25-1b of the text.

25-3C

(a) The low melting point of $ReOCl_4$ (30 °C) compared with that of $WOCl_4$ (209 °C) suggests a molecular rather than a network structure for the former. If bridging is as important a structural feature of the rhenium analog as it is for $WOCl_4$, the melting point of $ReOCl_4$ whould be much higher than it actually is.

(b) The structure, as shown in Figure 1, has an $ReCl_4$ pyramid, capped by an oxygen atom. These monomer units are then each bridged to another by weak Re····Cl interactions of two types: one to give dimers (Re type 2) and another to give infinite chains (Re type 1). These Re····Cl distances are very long: 3.55 and 3.65 Å, an indication that the $ReOCl_4$····$ReOCl_4$ interaction is extremely weak.

(c) The geometries suggested in the paper are the following:

Molecule	Gas-Phase Geometry
(i) $MoCl_5$	trigonal bipyramid
(ii) $WOCl_4$	square pyramid, basal chlorine
(iii) $WSCl_4$	square pyramid, basal chlorine
(iv) $ReOCl_4$	square pyramid, basal chlorine

A trigonal bipyramid is a reasonable (in fact the actual) structure for $MoCl_5$ in the vapor. By analogy to $ReOCl_4$, the author of this paper argues in favor of the square pyramid for $WOCl_4$ and $WSCl_4$.

CHAPTER 26

A - Review

26-1A
This is given in Table 26-1.

26-3A
Because the f orbitals shield one another poorly from the effects of
the growing nuclear charge on traversing the row from La to Lu, the
Lanthanides experience a dramatic increase in effective nuclear
charge with increase in atomic number. This causes a progressive
shrinking (termed the Lanthanide contraction) in the atomic and ionic
radii of the elements from La to Lu. In turn, subsequent elements
have dramatically smaller radii than might be otherwise expected
based solely on atomic number. See Figure 27-1, Table 26-1 and
Section 8-12.

26-5A
Scandium, although it resembles aluminum, is conveniently considered
here because it is the first member of Group IIIA(3), of which La is
a member. According to this reasoning, we might also take up
actinium in this chapter. Because Y^{3+} has a radius similar to Tb^{3+}
and Dy^{3+}, the lanthanide minerals always contain some percent of
yttrium. Also, the chemistry of Y (e.g. aqua ions, hydrolysis and
complexes) is similar to that of other lanthanides.

26-7A
The early members of the lanthanide series are relatively large ions
and may form 7-, 8-, 9-, or even 12-coordinate complexes. Latter
members of the lanthanides, being smaller, more commonly form 6-
coordinate complexes than do the larger ions.

26-9A
Divalent ions form sulfate precipitates. Eu^{2+} also forms an
insoluble carbonate. Trivalent lanthanides are readily and
characteristically precipitated as the fluorides. Also, the
hydroxides, $M(OH)_3$, form gelatinous precipitates. Cerium(IV) forms
the nonbasic CeO_2 as well as the basic hydrous oxide, $CeO_2 \cdot nH_2O$. The

latter is obtained as a gelatinous solid when Ce^{IV}(aq) is treated with OH^-(aq). Other insoluble compounds of Ce^{IV} include the phosphate, the iodate, and the oxalate.

26-11A

Volatile compounds of the lanthanides are given by some β-diketones, especially fluorinated ones. Such complexes are readily separated by gc techniques and the like. These are often useful as nmr shift reagents as outlined in Section 26-3.

B - Additional Exercises

26-1B

Ion	Electron Configuration	Number of Unpaired Electrons	S
Pr^{3+}	$[Xe]4f^2$	2	1
Pm^{3+}	$[Xe]4f^4$	4	2
Sm^{2+}	$[Xe]4f^6$	6	3
Gd^{3+}	$[Xe]4f^7$	7	7/2
Tb^{4+}	$[Xe]4f^7$	7	7/2
Tm^{3+}	$[Xe]4f^{12}$	2	1
Lu^{2+}	$[Xe]4f^{14}5d^1$	1	1/2

26-3B

(a) $Pr_2O_3 + 6NH_4Cl \rightarrow 2PrCl_3 + 3H_2O + 6NH_3$

(b) $2CeO_2 + Sn^{2+} + 8H^+ \rightarrow Sn^{4+} + 2Ce^{3+} + 4H_2O$

(c) $CeO_2 \cdot nH_2O + 4HCl \rightarrow [Ce(H_2O)_n]^{4+} + 2H_2O + 4Cl^-$

$\quad 4[Ce(H_2O)_n]^{4+} + 2H_2O \rightarrow 4[Ce(H_2O)_n]^{3+} + 4H^+ + O_2$

$\quad 2[Ce(H_2O)_n]^{4+} + 2Cl^- \rightarrow 2[Ce(H_2O)_n]^{3+} + Cl_2$

(d) $2Ce^{3+} + S_2O_8^{2-} \rightarrow 2Ce^{4+} + 2SO_4^{2-}$

26-5B

$$4e^- + 4Pr^{4+} \rightarrow 4Pr^{3+} \qquad E_{red} = 2.9 \text{ V}$$
$$2H_2O \rightarrow O_2 + 4H^+ + 4e^- \qquad E_{ox} = -1.2 \text{ V}$$

$$4Pr^{4+} + 2H_2O \rightarrow 4Pr^{3+} + O_2 + 4H^+ \qquad E_{cell} = 1.7 \text{ V}$$

C - Questions from the Literature of Inorganic Chemistry

26-1C

(a) $$Na + (CH_3)_2CHOH \rightarrow Na^+OCH(CH_3)_2{}^- + 1/2H_2$$
$$6NdCl_3 + 17NaOCH(CH_3)_2 \rightarrow Nd_6[OCH(CH_3)_2]_{17}Cl + 17NaCl$$

(b) The number of starting moles of $NdCl_3$ is:

$$4.6 \text{ g} \times 1 \text{ mole}/250.6 \text{ g} = 0.018 \text{ mole } NdCl_3$$

Therefore the number of moles of product to be expected is:

$$0.018 \text{ mole } NdCl_3 \times 1 \text{ mole product}/6 \text{ mole } NdCl_3 = 3.0 \times 10^{-3} \text{ mol}$$

The expected yield of product is, therefore:

$$3.0 \times 10^{-3} \text{ mole} \times 1905.6 \text{ g/mole} = 5.7 \text{ g}.$$

Based, however, on the amount of sodium available, the number of moles of product to be expected is:

$$.055 \text{ mole Na} \times 1 \text{ mole product}/17 \text{ mole Na} = 3.2 \times 10^{-3} \text{ mol}$$

The reaction is essentially quantitative, according to these calculations.

(c) The necessary information is given in Figure 1 of the article.

(i) The six terminal oxygen atoms are O1, O2, O3, O4, O5, and O6.

(ii) The edge-bridging oxygen atoms are of two classes: O7 through O12; and O13, O14, and O15.

(iii) There are two trigonal face-capping oxygen groups, O16 and O17.

(d) The Nd cluster is arranged as a trigonal prism.

(e) These may be regarded as Nd^{III} ions. The magnetic susceptibility data (μ_{eff} = 3.22 BM per Nd atom) suggests (*cf.* Table 2-4 of the text) the presence of three unpaired electrons on each Nd^{III} ion. The electron configuration may then be written straightforwardly: $[Xe]4f^3$.

A - Review

<u>27-1A</u>

See Table 27-1.

<u>27-3A</u>

We have thorium (^{232}Th) found in monazite sands and uranium (^{235}U, ^{238}U) found in various ores such as pitchblende and uraninite, which are different forms of the oxide, ranging in composition from UO_2 to U_3O_8. Traces of Pa are also available from uranium ores. All transuranium elements are made artificially in reactors by, for example, neutron capture as in equations 27-2.3 and 27-2.4. Those that are available in significant quantities are isotopes of Am, Cm, Bk, Cf, Es, and Fm, the latter in only μg amounts.

<u>27-5A</u>

The ion Cm^{3+} has the configuration $[Rn]5f^7$, as do Am^{2+} and Bk^{4+}.

<u>27-7A</u>

The metal halides or oxides are reduced by vapors of Li, Mg, or Ca at high temperatures. The bulk metals are silvery white, and react with oxygen, especially when finely powdered. They dissolve readily in acids such as HNO_3 and HCl. The most dense metals are U, Np, and Pu. Americium and Cm, on the other hand, are lighter metals.

<u>27-9A</u>

Actinium is available from uranium ores (*c.a.* 0.1 mg of actinium per ton of uranium ore) as a natural decay product of ^{235}U, but it is now most readily available from the spent fuel elements of nuclear reactors, where it is produced as in equation 27-4.1.

<u>27-11A</u>

Uranyl nitrate is precipitated as the hydrous oxide $UO_2(OH)_2 \cdot H_2O$ from aqueous ammonia. The hydrous oxide is heated to give UO_3. The oxide may be reduced by Li, Mg, Ca, or Ba, at 1400 °C, but Mg or Ca is normally used.

27-13A

The pyrophoric black powder UH_3 is obtained, even at 25 °C, by direct interaction of the elements, as in equation 27-7.3. Some uses are outlined in equations 27-7.4 through 27-7.8.

B - Additional Exercises

27-1B

These are listed in Section 27-8. First, the varying oxidation states are exploited to allow preferential extraction or precipitation of one ion over another. Second, the ions $MO_2{}^{2+}$, M^{4+}, and M^{3+} are each preferentially soluble (and hence extractable) into either ether or kerosene (with tributylphosphate). Third, the higher oxidation state ions (*i.e.* $MO_2{}^{2+}$) precipitate as the fluorides or the phosphates under proper conditions. Also we have ion exchange procedures. Two cycles are illustrated in Schemes 27-I and 27-II. The latter are used in the recovery of Pu or Np from spent fuel assemblies of nuclear reactors. The details are listed in Section 27-8.

27-3B

This is Scheme 27-II.

27-5B

This comparison is crucial to an understanding of Schemes 27-I and 27-II. Solid compounds are usually isomorphous. For uranium, $UO_2{}^+$ is not as stable as $UO_2{}^{2+}$. $NpO_2{}^+$ is stable. All four oxidation states for Pu have similar stabilities, and may coexist in 1M $HClO_4$. Americium is most stable as Am^{3+}, and only powerful oxidants give higher oxidation states. The solubilities in kerosene with 30 % TBP follow the trend $UO_2{}^{2+} > NpO_2{}^+ > PuO_2{}^{2+}$. The stabilities follow the trend $UO_2{}^{2+} > NpO_2{}^+ > PuO_2{}^{2+} > AmO_2{}^{2+}$, such that $AmO_2{}^{2+}$ is most easily reduced. The ions $MO_2{}^+$ disproportionate as in equation 27-3.1 most easily in the order Np < Am < Pu < U.

CHAPTER 27

C - Questions from the Literature of Inorganic Chemistry

27-1C

(a) The dicarbollide ion, $[C_2B_9H_{11}]^{2-}$ (which is shown in Figure 12-9 of the text), is a six-electron π donor (as is $C_5H_5^-$). However, the dicarbollide ion is larger and more negative than $C_5H_5^-$. Since actinide compounds contain larger ions with higher formal oxidation states, the dicarbollide anions should be well suited to actinide complex formation.

(b) Each lithium cation is tetrahedrally coordinated by four tetrahydrofuran (THF) molecules. See Figure 3.

(c) The starting material UCl_4 contains U^{IV}, as does the product.

(d) Actinide ions are larger and can accommodate more ligands than typical transition metals from the d-block of the periodic table. Here, the coordination number is, formally, eight.

(e) The formal coordination number is taken to be eight, since there are eight coordinated electron pairs: three each from the dicarbollide ligands and one each from the chloride ligands. Although this compound was not found to add more ligands, a coordinatively saturated actinide compound normally is taken to have coordination number ten. Apparently the coordinatively saturated ion $[U(C_2B_9H_{11})_3]^{2-}$ cannot be formed due to the bulk of the dicarbollide ligands. It might be an interesting experiment to discover if other monodentate ligands can be added.

27-3C

(a) $2UF_6 + SiO_2 \rightarrow 2UOF_4 + SiF_4$

$UF_6 + SiO_2 \rightarrow UO_2F_2 + SiF_4$

(b) $4HF + SiO_2 \rightarrow SiF_4 + 2H_2O$

(c) Total hydrolysis is accomplished with an excess of water and gives, violently, the oxide and HF. Controlled hydrolyses lead to the intermediate products UOF_4 and, subsequently, UO_2F_2.

(d) The uncertainty here arises because of the indistinguishable nature of F atoms and O atoms based on the X-ray diffraction data. Arguments favoring an axial oxygen atom are the following. The vibrational spectrum shows a single, strong U-O band that is not characteristic of U-O-U linkages. Thus, the oxygen is terminal. Second, all known U-O bonds are short and terminal. Third, the bridging distances (2.25 - 2.29 Å) found here in UOF_4 are typical of other known fluorine bridges (2.24 - 2.43 Å).

(e) There are quite a few uranium-fluorine-uranium bridging interactions, as shown in Figure 3 and 4. In fact, only the axial fluorine atom (either $X_{ax(1)}$ or $X_{ax(2)}$) and one equatorial fluorine atom (F_3 in Figure 3) are nonbridging. As shown in Figure 4, two equatorial fluorine atoms per uranium atom are involved in bridging to another uranium atom. Thus, we have three classes of fluorine atoms in UOF_4: axial fluorine atoms, nonbridging equatorial fluorine atoms (F_3), and two equatorial bridging fluorine atoms.

A - Review

<u>28-1A</u>

First, we have π-acceptor ligands that use empty π^* orbitals
analogous to the π_2 levels in Figure 3-27. These are: CO, N_2RNC,
CN^-, NO, and CS. The cyclic, aromatic ligands pyridine and 1,10-
phenanthroline accept metal electron density into ring π^* orbitals.
The ligands with Group VB(15) donor atoms that are π acceptors are,
for example, the trialkyls PR_3 (3d), AsR_3 (4d), and SbR_3 (5d). The
ligands SR_2 are weak π acceptors, utilizing the 3d orbitals of
sulfur.

<u>28-3A</u>

The simplest carbonyls of metals such as Mn and Co are polynuclear
because the metal atom has an odd number of electrons. This requires
sharing of metal electrons in a metal-metal bond in order to satisfy
the noble gas formalism (effective atomic number rule). Note that we
arrive at this conclusion regardless of whether or not the
metal-metal bond is supported by bridging carbonyl groups.

<u>28-5A</u>

There are bridging and terminal CO ligands, as shown in Structures
28-Ia and 28-Ib, respectively. Additionally, there are the face-
capping, or triply-bridging CO ligands depicted in Figure 28-1.
There are also numerous examples of the so-called semi-bridging
carbonyl ligands, which may be diagrammed generally as follows:

although there are wide variations in the precise geometry of the
semi-bridging carbonyl group from one compound to another. Semi-
bridging carbonyl ligands are to be regarded as four-electron donors,
one σ pair to M_1 and one π pair to M_2. For more details see (a) F.
A. Cotton and G. Wilkinson, "Advanced Inorganic Chemistry," Fourth

Edition, Wiley Interscience, New York, 1980, Section 25-4, and (b) C. M. Lukehart, "Fundamental Transition Metal Organometallic Chemistry," Brooks/Cole, Monterey, CA, 1985.

28-7A

These are shown in Figure 28-2. Note the presence of bridging CO ligands in $Fe_3(CO)_{12}$ and the lack thereof in the Ru and Os analogs.

28-9A

Metal compounds (typically oxides, carbonates or chlorides, *etc.*) are reduced in the presence of excess CO. The reducing agents are CO itself, or Na, Al, Mg, Cu and the like. See especially equations 28-5.1 through 28-5.4. Another useful reducing agent is the ketyl radical anion of benzophenone, Ph_2CO^-, prepared in anhydrous THF by reduction of benzophenone with sodium. Further examples are listed in the answer to question 28-10A.

28-11A

There are two geometrical isomers. The *trans* isomer has one CO stretching band, whereas the *cis* isomer has four.

28-13A

(a) Since triethylamine is not a π acceptor, the CO group *trans* to it has no competition for π-electron density, and the extent of π back-bonding to this one CO group is higher than would be true of the average extent of π back-bonding to a set of mutually *trans* CO ligands. We therefore expect a decrease in the CO stretching frequency, because the CO $\pi*$ orbital is more populated when the CO ligand is *trans* to triethylamine.

(b) Less π bonding occurs overall when a positive charge is introduced on the metal. Therefore, the CO stretching frequence should increase.

(c) The negative charge increases the amount of electron density available for π back-bonding, and we expect the CO stretching frequency to decrease, because the CO $\pi*$ orbital is more populated.

28-15A

In complexes containing the linear M-NO group, the nitrogen monoxide (nitric oxide) ligand is best regarded as a three electron donor, although the formal result is the same as that for CO ligands; there is a σ donor pair and a π acceptor pair of electrons involved in the metal-ligand bond. Formally, NO^+ is isoelectronic with CO, and we may consider that an extra electron of NO is allocated to the metal, which then has a formal oxidation state one lower than if the NO ligand is counted as a three electron donor. Whether we say that NO is a two electron σ donor and a one electron π-acceptor, or that NO^+ is a two electron σ donor and a two electron π-acceptor is a matter of formality.

28-17A

Nitrogen monoxide (nitric oxide) may be counted as a three electron donor ligand if one σ pair is given to the metal in the usual fashion, and if the metal contributes one electron, along with the odd electron already on NO, to form a metal-to-ligand π back-bond analogous to that of a M-CO group. Conversely, we may begin by assigning the odd electron of NO to the metal completely, giving, in a formal sense, NO^+, which is straightforwardly isoelectronic with CO.

If each of the bidentate or η^2 ligands depicted in this problem is considered to be neutral, then they should be counted as three electron donors. If each is formally taken to be anionic (perhaps a more reasonable approach), then the three ligands shown in the problem are four electron donors. The stoichiometries and the formal oxidation states of the metals should be different from those of NO complexes.

28-19A

The series shown in Section 28-15 indicates that PF_3 is most similar to CO.

B - Additional Exercises

28-1B

Formation of the dimer $V_2(CO)_{12}$ ($c.f.$ $Mn_2(CO)_{10}$ and $Co_2(CO)_8$) requires that each vanadium atom become seven coordinate, a situation that is, perhaps, unreasonably congested. Although preparation of the dimer has been claimed (Ford, T. A.; Huber, M.; Klotzbucher, W. K.; Moskovitz, M.; and Ozin, G. A., $Inorg.$ $Chem.$ **1976**, 15, 1666), the paramagnetic $V(CO)_6$ (μ_{eff} = 1.81 BM) most readily achieves the noble gas configuration by reduction to the anion, $[V(CO)_6]^-$.

28-3B

The d orbitals of the lanthanide metals are less available for back bonding to CO ligands than are the d orbitals of the transition metals of the d-block of the periodic table. Consequently, it is rare that the lanthanides form simple carbonyl derivatives. The organometallic chemistry of the lanthanides most typically involves ionic derivatives, for instance, with the cyclopentadienyl anion, $C_5H_5^-$. CO and other ligands such as tetrahydrofuran are readily added to lanthanide compounds in order to satisfy coordination number requirements. It is also common to find CO behaving, like THF, as an O-donor ligand.

28-5B

The corresponding formula may be generated in two different ways: (a) two NO ligands at the same metal may be replaced by three CO ligands, leading to an increase in coordination number or (b) one NO may replace one CO if the metal is changed to one having the next lower atomic number.

(a) $Cr(NO)_4$, $Cr(CO)_6$, $V(CO)_5(NO)$
(b) $Mn(CO)(NO)_3$, $Mn_2(CO)_{10}$, $Cr_2(CO)_8(NO)_2$
(c) $Mn(CO)_4(NO)$, $Mn_2(CO)_{10}$, $V_2(CO)_6(NO)_4$
(d) $Co(NO)_3$, $Co_2(CO)_8$, $Fe_2(CO)_6(NO)_2$
(e) $Fe(CO)_2(NO)_2$, $Fe(CO)_5$, $Mn(CO)_4(NO)$

28-7B

Ligands such that are CO analogs, such as RNC, are σ-donor ligands
and π-acceptor ligands. The alkenes, as discussed in Chapter 29, are
π donors and π acceptors. Also, the alkene ligands bond to metal
atoms in a side-on fashion.

28-9B

A proton more readily ionizes from $H_2Fe(CO)_4$ than from $[HFe(CO)_4]^-$.
In fact, whereas $H_2Fe(CO)_4$ is acidic, $[HFe(CO)_4]^-$ is hydridic. Both
hydrogen atoms of $H_2Fe(CO)_4$ are obviously bound to the iron atom.
Note that whereas $H_2Fe(CO)_4$ is relatively stable, the protonation of
$[HCr(CO)_5]^-$ leads to loss of H_2 and formation of $[Cr(CO)_5]$.
$[HCr(CO)_5]^-$ is more hydridic than $[HFe(CO)_4]^-$.

28-11B

Only $V(CO)_6$, "$[Cr(CO)_5]^-$" and "$HMn(CO)_4$" do not obey the 18-electron
rule. The proper substances are $[Cr(CO)_5]^{2-}$ and $HMn(CO)_5$. Both
$V(CO)_6$ and $[V(CO)_6]^-$ are known.

28-13B

$V(CO)_6$ is a 17-electron compound.

28-15B

See Figure 3-37 and the discussion at the end of Section 3-7. The sp
hybrid orbital of the CO carbon atom overlaps with a hybrid from each
metal, forming a four-center, two-electron bond. The CO stretching
frequencies for such triply-bridging CO ligands are lower than those
of doubly-bridging CO ligands.

28-17B

(a) $4Ru^{3+} + 20NH_3 + N_2H_4 + 4OH^- \rightarrow [Ru(NH_3)_5N_2]^{2+} +$
$$3[Ru(NH_3)_5OH_2]^{2+} + H_2O$$

(b) $[Ru(NH_3)_5OH_2]^{3+} + N_3^- \rightarrow [Ru(NH_3)_5N_2]^{2+} + H_2O + 1/2N_2$

(c) $[Ru(NH_3)_5OH_2]^{2+} + N_2 = [Ru(NH_3)_5N_2]^{2+} + H_2O$

(d) $RuCl_3 + 5NH_3 \rightarrow [Ru(NH_3)_5Cl]^{2+} + 2Cl^-$

$2[Ru(NH_3)_5Cl]^{2+}(aq) + Zn \rightarrow 2[Ru(NH_3)_5OH_2]^{2+} + ZnCl_2(aq)$

$[Ru(NH_3)_5OH_2]^{2+} + N_2 \rightarrow [Ru(NH_3)_5N_2]^{2+} + H_2O$

C - Questions from the Literature of Inorganic Chemistry

28-1C

(a) $PPN^+[W(CO)_5Cl]^- + CH_3OH + PMe_3 \rightarrow W(CO)_5PMe_3 + PPN^+CH_3O^- + HCl$

The synthesis referenced in this paper employed a Lewis acid catalyst for the substitution of the chloride ligand in the starting materials. It is likely that methanol serves this function here.

(b) For each of the compounds mentioned in the introduction of this paper, there is some steric problem that makes the P-M bond longer than would be expected in the absence of such steric effects.

(c) First, the molecule is oriented in the unit cell in an uncongested way, as shown in Figure 1. Second, no intermolecular contact is less than the sum of the appropriate van der Waals radii, meaning that the intramolecular distances are unaltered by intermolecular interactions. Third, the ligands are arranged in a nearly perfect octahedral fashion around the tungsten atom, and the C-O bond lengths are nearly all the same. Last, the trimethylphosphine ligand is not distorted in any way from ideal geometry.

(d) The W-P distance in $W(CO)_5[P(t-Bu)_3]$ is longer than that in $W(CO)_5PMe_3$ by 0.170 Å, due to steric reasons. A greater lability for the $P(t-Bu)_3$ ligand would arise if steric acceleration operated. This is typical of a dissociative substitution mechanism.

28-3C

(a) The THF ligands are neutral, two-electron donors. The phenyl monoanions are two-electron, η^1, σ donors. The metal is Cr^{III}.

(b) The presence of Cr-to-phenyl π-back bonding is indicated by the short Cr-C distances.

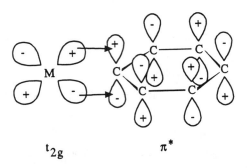

t_{2g} π^*

(c) Neither this nor the title compound satisfies the formalism.

28-5C

(a) In both Fe_2CO_9 and μ-$H[Fe_2(CO)_8]^-$, there are two nearly octahedral iron atoms bridged by three ligands. Thus, the dimers may be said to be constructed of two octahedra sharing a trigonal face. Carbonyl ligands bridge the iron atoms in $Fe_2(CO)_9$. Two carbonyl ligands and one hydride ligand bridge the two iron atoms in $[HFe_2(CO)_8]^-$.

(b) The three M-L-M bridge bonds in $[HFe_2(CO)_8]^-$ may be regarded as three-center, two-electron bonds, as shown in structures I and II of the paper.

(c) First, the metal-metal distance in the anion $[HFe_2(CO)_8]^-$ is the same as that in $Fe_2(CO)_9$. In contrast, compounds in which Cl^- are introduced (structure III and footnote 14 in the article) have M-M distances significantly longer than those with corresponding μ-H bridges. Second, compounds with three-center, two electron bridging CO or H^- ligands have acute M-L-M bridge angles, whereas the organic carbonyls or the chloro-bridged M-L-M systems typically have much larger angles. Finally, the systems with three-center, two-electron M-H-M or M-C(O)-M bridge bonds are fluxional.

(d) The presence of bridging CO ligands is suggested by the low CO stretching frequencies (1790 and 1750 cm^{-1}) in the infrared spectrum of the compound.

CHAPTER 29

A - Review

29-1A

An organometallic compound is one that contains a bond between a
metal atom and one or more carbon atoms of an organic ligand.

29-3A

These are listed in Section 29-2. (1) Alkyl or aryl halides may be
reacted (as in equation 29-2.1) directly with the more active metals:
Mg, Li, Na, K, Ca, Zn, and Cd. This is the principal method used to
prepare lithium alkyls, Section 29-3. (2) Halide compounds may be
alkylated with the alkyl reagents such as LiR, as well as with
aluminum alkyls or sodium cyclopentadienide. Reactions 29-2.2
through 29-2.4 serve as examples. (3) Metal hydrides react with
unsaturated organic compounds to give metal alkyls, as in reaction
29-2.8. This constitutes insertion of an olefin into a
metal-hydrogen bond. (4) Oxidative addition of an alkyl halide to an
unsaturated metal complex generates a metal-carbon bond, an example
being reaction 29-2.9.

29-5A

Chelating ligands such as TMED enhance the reactivity of the lithium
alkyls by promoting the formation of monomeric lithium alkyls in
solution.

29-7A

A Grignard reagent in ether solution generally enters into the type
of equilibrium shown in equation 29-5.2. Each compound involved in
this equilibrium is solvated, and both halide-bridged and alkyl-
bridged structures are involved, although the latter predominate only
for the methyl derivatives.

29-9A

Methyl lithium has the tetrameric structure shown in Figure 29-1.
The dietherate of methylmagnesium bromide has a tetrahedral
arrangement of four groups around a central magnesium ion. Dimeric
structures with either bridging halide or methyl groups are also

possible in solution. The dialkyls of mercury have linear, molecular structures. The dimeric structure of "AlMe$_3$" is shown in Figure 29-2. The compound Me$_3$SnF is the polymer shown in Structure 29-IV, in which the SnMe$_3$ moiety is roughly planar.

29-11A

(a) $BCl_3 + 3PhMgCl \rightarrow BPh_3 + 3MgCl_2$
$BPh_3 + PhMgCl \rightarrow BPh_4^- + Mg^{2+} + Cl^-$
$Na^+ + BPh_4^- \rightarrow NaBPh_4$

(b) $Hg + C_3H_5Br \rightarrow C_3H_5HgBr$

(c) $Et_2Hg + Zn \rightarrow Et_2Zn + Hg$

(d) $2Al + 3Me_2Hg \rightarrow [Me_3Al]_2 + 3Hg$

29-13A

A strong base such as butyllithium deprotonates the quaternary methylphosphorus compound as in reaction 29-10.4 to produce a Wittig reagent. This reacts with aldehydes or ketones to give olefins as in reaction 29-10.5.

29-15A

The transfer of a hydrogen atom from the beta carbon atom to the metal atom is accompanied by elimination of an olefin. This general reaction is termed β elimination, and it leads to the ready decomposition of most metal alkyls, as in the following example:

$$[EtW(CO)_5]^- \rightarrow [HW(CO)_5]^- + C_2H_4$$

29-17A

The structure of the anion in Zeise's salt is shown in Figure 29-3A. The salient structural features of this coordination ion, which are adequately explained by the bonding scheme of Figure 29-4, are that the C=C axis lies perpendicular to the [PtCl$_3$] plane and that the bond axis from the Pt atom to the olefin bisects the C=C distance.

29-19A

The iron and the platinum compounds have Structures 29-XIII and

29-XIV, respectively. The chromium complex is similar to that shown in Structure 29-XII, having an η^6-triolefinic attachment of COT, which is a six-electron donor ligand.

29-21A

These are equations 29-14.1 and 29-14.2.

29-23A

(a) For the carbenes, the metal is in a low oxidation state, whereas for alkylidenes the metal typically has a high oxidation state. Whereas carbene ligands may be regarded as σ-donor and weak π-acceptor ligands, the alkylidene ligand is properly regarded as being both a σ-donor and a π-donor ligand. Carbene carbon atoms are attacked by nucleophiles, whereas alkylidene carbon atoms are attacked by electrophiles. Carbene ligands typically have nonmetal substituents (as in Structures 29-XXXI and 29-XXXII) with heteroatoms that stabilize the electron-deficient, sp^2-hybridized carbene carbon atoms by π donation. The substituents on the typical alkylidene carbon atom are hydrogen or alkyl groups. Hence, the extent of M=C double bonding is greater in the alkylidenes than in carbenes.

(b) These compounds contain what is, formally, a M≡C-R moiety. We distinguish between complexes with a metal in a low oxidation state (carbynes) and those with a metal in a high oxidation state (alkylidynes).

29-25A

Attack of a nucleophile at the carbon atom of a coordinated CO ligand, followed by electrophilic attack at oxygen by R_3O^+, as in reaction 29-17.1, gives a carbene complex.

B - Additional Exercises

29-1B

Mercurinium ions such as those shown in Structures 29-I and 29-II, and in reaction 29-6.8, are cyclic intermediates in mercuration or oxomercuration reactions. The synthetic goal is the conversion of an alkene to an alcohol, as in reaction 29-6.9.

29-3B

The pseudo-tetrahedral, quaternized phosphorus atom in the cation $[Ph_3P-CH_3]^+$ is first deprotonated as in equation 29-10.4, using the strong base n-butyl$^-$. This produces butane, LiBr, and a phosphorane alkylidene, $Ph_3P=CH_2$. The latter reacts as in equation 29-10.5 with the C=O double bonds of ketones or aldehydes to give Zwitterions such as 29-V. Elimination of the stable Ph_3PO, triphenyl phosphine oxide, gives alkenes.

29-5B

This question is fully addressed in Section 30-6. Note that it is Markovnikov addition of the metal hydride across the double bond (as in reaction 30-6.5) that allows for isomerization to the hex-2-ene, and that this may give *cis* or *trans* olefins because of the free rotation about C-C single bonds.

(a) The key steps in the sequence are (1) coordination of the olefin to the metal center (reaction 30-6.1), (2) transfer of the hydride ligand to the olefin, giving a metal alkyl (reaction 30-6.2), and (3) β elimination to give an olefin and a metal hydride (the reverse of reaction 30-6.2). The *cis* and *trans* isomers of 2-hexene arise because the reverse of reaction 30-6.5 may be accomplished by β elimination of either one of the two hydrogen atoms on the "CH_2R" β carbon of (B) in equation 30-6.5.

(b) Incorporation of D from L_nMD may proceed as in equations 30-6.4 or 30-6.5, and D can end up (assuming only limited isomerization of the hex-2-ene products noted above) on any carbon of the original double bond.

29-7B

The dianion, $[Fe(CO)_4]^{2-}$, is nucleophilic and displaces I$^-$ from ICH_2NMe_2:

$$[Fe(CO)_4]^{2-} + ICH_2NMe_2 \rightarrow [(CO)_4Fe-CH_2NMe_2]^- + I^-$$

Transfer of a hydride group between two of the monoanions produced in the above reaction:

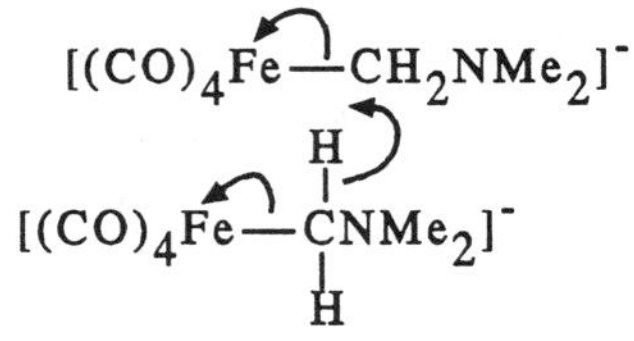

$$[(CO)_4Fe-CH_2NMe_2]^-$$

$$[(CO)_4Fe-\overset{\overset{\displaystyle H}{|}}{\underset{\underset{\displaystyle H}{|}}{C}}NMe_2]^-$$

gives trimethyl amine, $[Fe(CO)_4]^{2-}$ and the carbene shown below.

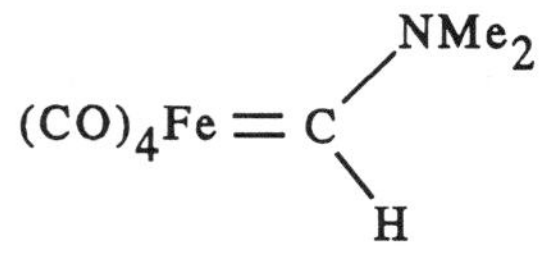

$$(CO)_4Fe = C \overset{\displaystyle NMe_2}{\underset{\displaystyle H}{<}}$$

29-9B

In both cases, the ligand is attached "side-on". The bonding model for the olefin is depicted in Figure 29-4. There is an entirely analogous situation in the Pt-O_2 "side-on" bond. These bonds are therefore constructed of two components. First, we have donation of pπ-pπ bonding electron density of either C_2H_4 or O_2 into an empty metal orbital (left side of Figure 29-4). Second, there is back donation from a filled metal dπ orbital into the empty π* orbital of either O_2 or C_2H_4 (right side of Figure 29-4). Each component of this bond system weakens the intra ligand bond. Consequently, the C-C bond length of the olefin increases by about 0.15 Å upon coordination, and the O-O distance increases from 1.21 Å (O_2) to 1.45 Å (Pt-O_2).

29-11B

In the iron butadiene reactant we have an η^4-butadiene ligand that is a 4-electron donor:

η^4-$C_4H_6Fe(CO)_3$

butadiene,	4 electrons
iron(0),	8 electrons
3 CO,	6 electrons
total	18 electrons

In the product of reaction 29-16.4, we have an η^3-allylmethyl ligand which should be regarded as a three electron donor (with an iron(I) oxidation state) or as a four electron donor (with an iron(II) oxidation state):

$(\eta^3$-allylmethyl$)Fe(CO)_3Cl$

η^3-allylmethyl,	3 electrons
chloride,	2 electrons
3 CO,	6 electrons
iron(I),	7 electrons
total	18 electrons

29-13B

$Mo(CO)_6$ (Table 28-1) has the octahedral structure. C_7H_8 is cyclohepta-1,3,5-triene, whose structure is shown as the molybdenum complex in Structure 29-XII. The dimeric compound, $[(C_2H_4)_2RhCl]_2$, is likely analogous to the dimer shown in Figure 25-9.

29-15B

Zeise's salt is both coordinatively and electronically unsaturated, and should readily add more ligands.

29-17B

(a) $BCl_3 + 4RNa \rightarrow Na[BR_4] + 3NaCl$

(b) $Et_3Al + EtLi \rightarrow Li[AlEt_4]$

(c) $Mg + EtBr \rightarrow EtMgBr$

(d) $PCl_3 + 3EtMgBr \rightarrow PEt_3 + 3MgBrCl$

C - Questions from the Literature of Inorganic Chemistry

29-1C

(a) The Dewar-Chatt-Duncanson bonding model that is mentioned in the article is essentially that which is diagrammed in Figure 29-4 of the text. The ligand with the shortest Fe-C bond distances is the one in which the olefin is the strongest π acceptor, because, as pointed out in the article, it is the π back bond (shown on the right of Figure 29-4) that is the most important component of the iron-olefin bond. Hence, the loss of planarity of the alkene upon coordination and the various lengths of bonds (C=C and Fe-C) ought to reflect the extent of π back bonding in these two complexes. As discussed by the authors, and as illustrated by the data of Table VIII in this article, it is the DEM ligand that is the stronger π acceptor. This is shown also by the infrared and structural parameters regarding the equatorial CO ligands, which indicate that DEM is nearly as good a π acceptor as CO itself.

(b) In each case we have an η^2-olefinic ligand that is a two-electron donor, three carbonyl ligands (2 x 3 = 6 electrons) a triphenylphosphine ligand (2 electron donor) and iron(O), a d^8 metal. The total is 18 electrons in each case.

(c) Olefins such as these prefer equatorial sites on trigonal bipyramidal complexes because of electronic reasons; this site provides the greatest extent of metal-olefin π back bonding. Note that the authors argue that the phosphine ought to prefer an axial coordination site, but that in the DEF complex shown in Figure 3 of the article, this is prevented for steric reasons.

(d) The iron-to-C bond lengths of the CO ligands reflect the competition with other π back-bonding ligands for π electron density from the central metal. A ligand that effectively competes for π electron density with the CO ligand weakens the Fe-CO bond by reducing the strength of the π back bond to the CO ligand. Thus, the Fe-CO bond distances (and the ν_{CO} stretching frequencies from the infrared spectra) can be used as a measure of another ligand's success at competing with CO ligands for π-electron density on the metal. The longer the Fe-CO bond length, the better are the π-acceptor capabilities of the ligand that competes with a CO group for π-electron density on the metal. The data of Table VI in this article may therefore be used to establish the following order of π acidities: CO > PPh$_3$, and DEM > MA. The steric effects and the different coordination positions adopted by DEM and DEF make it difficult to use Fe-CO data in this fashion to judge the relative π acidities of DEM and DEF. For this we must reason as in the answer to part (a) above.

29-3C

(a) Re(CO)$_2$Cp(RC≡CR):

RC≡CR,	2 electrons
2 CO,	4 electrons
Cp$^-$,	6 electrons
Re^I,	6 electrons
total	18 electrons

(b) The important differences are set down in Table II of this article, where the d^6 rhenium compound is seen to have a diphenylacetylene ligand that is less distorted upon coordination to a metal than are most of the other alkyne ligands in the list. This ligand is an η^2 two-electron donor. By contrast, the diphenylacetylene ligand of the Ta complex found in question 2 above is an η^2 four-electron donor, and it is strongly distorted by coordination to the metal atom. The one obvious difference between these two different complexes is the number of electrons on the two metal atoms.

A - Review

30-1A

As explained in Section 30-1, a coordinatively unsaturated substance
is one that may expand its coordination number by the addition of
more ligands. This property is a necessary condition for activity as
a catalyst. The addition of ligands to such a metal center requires,
in some cases, the displacement of solvent molecules, and in other
cases, a change in oxidation state at the metal. When substitution
reactions are slow at room temperature, either thermal or
photochemical activation is necessary in order to accomplish these
sorts of ligand additions. Examples of these processes are shown in
equations 30-1.1 and 30-1.2, as well as 30-6.1 and 30-8.4.

30-3A

The oxidative additions of HCl, O_2, and H_2 to $Ir^ICl(CO)(PPh_3)_2$ are
accomplished as in reactions 30-2.15, 30-2.16, and 30-2.24,
respectively. The reaction with $(CF_3)_2CO$ proceeds in a manner
analogous to equation 30-2.17. Addition of CH_3I is potentially an
ionic process (equations 30-2.20 and 30-2.21) or a concerted one,
requiring nonpolar conditions, as in reaction 30-2.24. The latter
gives *cis* stereochemistry. The likely structures for the other two
addends are the following:

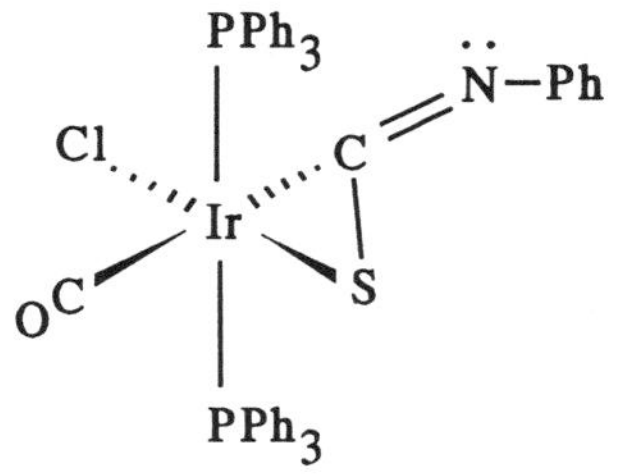

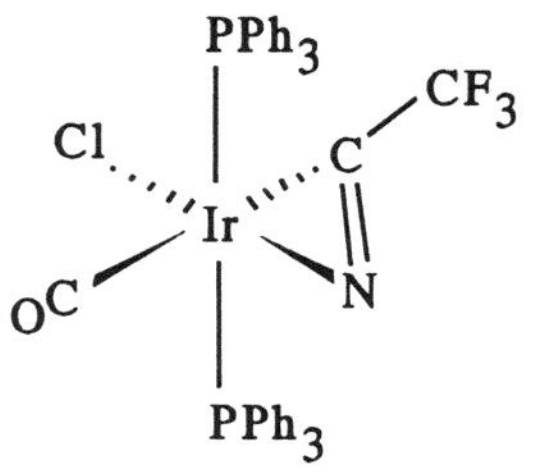

30-5A

The insertion reaction is shown in general terms in equation 30-3.1.
It is a reaction in which some group is inserted into a metal-ligand
bond. It is sometimes best regarded as migratory insertion, as
discussed for CO in the text. Examples of insertion reactions are
equations 30-3.2 through 30-3.8.

198

30-7A

(a) This is reaction 30-2.4. The protonated ruthenium complex cation product is octahedral, and it contains a Ru–H bond.

(b) This is reaction 30-4.3. The trigonal bipyramidal product contains the Ir–C(=O)–OCH$_3$ unit.

(c) This is attack by the methyl carbanion as in equation 30-4.4 to give an intermediate acyl ligand, and electrophilic addition of Me$^+$ using Me$_3$O$^+$, as in reaction 29-17.1. The product is an alkoxy carbene, analogous to structure 29-XXXI, or more precisely, analogous to the product of reaction 29-17.1.

(d) This is reaction 30-4.1, in which two equivalents of OH$^-$ convert a Fe–NO group into a Fe–NO$_2$ group.

(e) Nucleophilic attack gives:

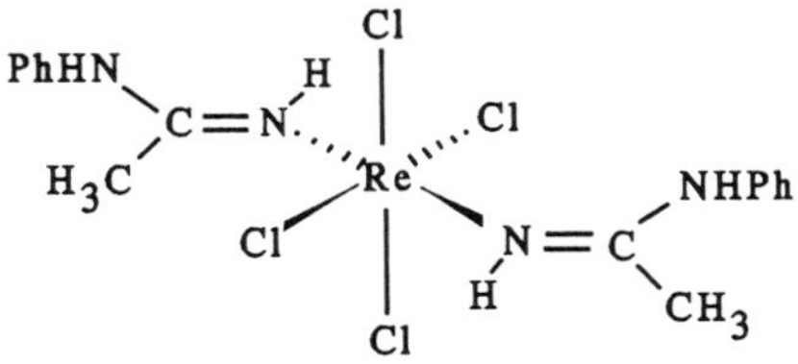

For more information, see the papers by G. Rouschias and G. Wilkinson, *J. Chem. Soc.*, **1968**, 489, and H. C. Clark and L. E. Manzer, *Inorg. Chem.*, **1971**, *10*, 2699.

30-9A

(a) Hydroformylation:

$$RCH=CH_2 + H_2 + CO \rightarrow RCH_2CH_2CHO$$

(b) Hydrosilylation:

$$RCH=CH_2 + HSiR_3 \rightarrow RCH_2CH_2SiR_3$$

(c) The Ziegler-Natta process:

$$nCH_2=CH_2 \rightarrow (CH_2CH_2)_n$$

(d) The Wacker process for acetaldehyde:

$$C_2H_4 + 1/2\ O_2 \rightarrow CH_3CHO$$

30-11A

It is convenient to list the copper reactions first. Thus, Cu^{II} chloride oxidizes Pd^0 as in equation 30-11.2 and oxygen oxidizes Cu^I chloride as in equation 30-11.3. These reactions thereby regenerate the Pd^{II} catalyst precursor. Addition of Cl^- to $PdCl_2$ gives the complex anion, $[PdCl_4]^{2-}$, which is the likely active catalyst. The next steps in the sequence are addition of the olefin (equation 30-11.5), and addition of water to give the neutral product of reaction 30-11.6. Further attack by water gives the intermediate shown in Structure 30-III. Elimination of water gives a hydride of Pd (reaction 30-11.7), which subsequently eliminates aldehyde according to reaction 30-11.8.

B - Additional Exercises

30-1B

The tetrakis(diethylamido)titanium(IV) complex is coordinatively unsaturated, and may add the CS_2 ligand:

$$Ti(NEt_2)_4 + CS_2 \rightarrow (Et_2N)_4Ti\text{-}SCS$$

This may rearrange via. a four-centered transition state:

$$(Et_2N)_3Ti\text{---}NEt_2 \longrightarrow (Et_2N)_3Ti\begin{array}{c}S\\ \diagup \diagdown \\ \diagdown \diagup \\ S\end{array}CNR_2$$

Such a sequence may be accomplished a total of four times.

30-3B

This is a solvated catalyst, and the two methanol ligands are readily substituted. The first step is oxidative addition of H_2:

$$[Rh^I(PEtPh_2)_2\ 2MeOH]^+ \rightarrow [Rh^{III}(PEtPh_2)_2(H)_2\ 2MeOH]^+$$

Next we have substitution of a methanol ligand by the incoming alkene to give an alkene-dihydrido rhodium complex similar to intermediate (B) in Figure 30-1. The last steps are insertion of the alkene ligand into a Rh-H bond to give a hydrido-alkyl rhodium complex, followed by elimination of the alkane.

30-5B

Some reasonable steps are the following. First we might have oxidative addition of HCN to the Ni^o center:

$$[P(OEt)_3]_4Ni^o + HCN \rightarrow [P(OEt)_3]_4Ni^{II}(H)(CN)$$

Then addition of the alkene and insertion of it into the Ni-H bond could give an alkyl-cyano nickel complex:

Transfer of the cyanide and elimination as follows gives a monocyano product:

Similar consequences for the alkene at the other end of the molecule give the product, 1,4-dicyanobutane.

30-7B

This type of migratory insertion reaction proceeds by the three-centered transition state shown in 30-3.11, and no change in the stereochemistry at the alkyl group is involved.

30-9B

Displacement of a triphenylphosphine ligand by ethylene and insertion of the ethylene into the Rh-H bond gives a Rh^I ethyl complex:

$$RhH(CO)(PPh_3)_3 + C_2H_4 \rightarrow Rh(C_2H_5)(CO)(PPh_3)_2 + PPh_3$$

Next, we have oxidative addition of benzoyl chloride to give an ethyl-benzoyl-chloro complex of Rh^{III}:

$$Rh(C_2H_5)(CO)(PPh_3)_2 \quad + \quad PhC\!\!\underset{Cl}{\overset{O}{\diagup}} \quad \longrightarrow$$

The latter undergoes reductive elimination of propiophenone, generating $RhCl(CO)(PPh_3)_2$.

30-11B

This fine work was reported by J. R. Sweet and William A. G. Graham, *J. Amer. Chem. Soc.*, **1982**, *104*, 2811-2815. The obvious importance regarding Fischer-Tropsch chemistry is that Graham has clearly demonstrated, first of all, the stepwise reduction of a coordinated (and activated) CO group, and secondly, the intermediacy in this process of formyl and hydroxymethyl complexes.

$[CpRe(CO)_2(NO)]^+$

Cp^-,	6 electrons
Re^{II},	5 electrons
NO,	3 electrons
2 x CO,	4 electrons
total	18 electrons

$CpRe(CO)(NO)-C(=O)H$

Cp^-,	6 electrons
Re^{II},	5 electrons
NO,	3 electrons
CO,	2 electrons
$-C(=O)H^-$,	2 electrons
total	18 electrons

$CpRe(CO)(NO)-CH_2OH$

Cp^-,	6 electrons
Re^{II},	5 electrons
NO,	3 electrons
CO,	2 electrons
$-CH_2OH$,	2 electrons
total	18 electrons

$CpRe(CO)(NO)-CH_3$

Cp^-,	6 electrons
Re^{II},	5 electrons
NO,	3 electrons
CO,	2 electrons
CH_3^-,	2 electrons
total	18 electrons

C - Questions from the Literature of Inorganic Chemistry

<u>30-1C</u>

(a) These are shown in the article, page 2668, where the oxidation states of the rhodium atoms are also listed.

(b) The various steps are the following, listed with numbers to identify the steps in the subsequent discussions. The first step is dissociation of a triphenylphosphine ligand to give the coordinatively unsaturated and active catalyst:

$$Rh(CO)(H)(PPh_3)_3 = Rh(CO)(H)(PPh_3)_2 + PPh_3 \qquad (1)$$

Next, we have coordination of the olefin in an equilibrium characterized by the equilibrium constant K_1, and giving a hydrido alkene of Rh^I:

$$Rh(CO)(H)(PPh_3)_2 + RCH=CH_2 = Rh(CO)(H)(PPh_3)_2(RCH=CH_2) \qquad (2)$$

This is followed by insertion of the alkene ligand into the Rh-H bond to give a square alkyl of Rh^I:

$$Rh(CO)(H)(PPh_3)_2(RCH=CH_2) = Rh(CO)(PPh_3)_2(RCH_2CH_2) \qquad (3)$$

The fourth step is the rate-determining step, namely oxidative addition of H_2, characterized by rate constant k_1, to give a coordinatively saturated dihydrido alkyl of Rh^{III}:

$$Rh(CO)(PPh_3)_2(RCH_2CH_2) + H_2 = Rh(CO)(H)_2(PPh_3)_2(RCH_2CH_2) \qquad (4)$$

Step number five is reductive elimination of an alkane, regenerating the active catalyst:

$$Rh(CO)(PPh_3)_2(H)_2(RCH_2CH_2) = Rh(CO)(H)(PPh_3)_2 + RCH_2CH_3 \qquad (5)$$

(c) There are a number of facts and reasonable proposals presented in this paper. Some of the pertinent ones are the following:

(i) Lower concentrations of the catalyst give apparently higher catalytic activity and lower catalyst selectivity. (See Figure 1 of the article.) This suggests that step (1) above, namely dissociation of one or more triphenylphosphine ligands, leads to a more active and less sterically encumbered metal center.

(ii) An increase in the partial pressure of hydrogen produces a corresponding increase in the rate of catalysis, as shown by the data of Figure 3. This is reasonable if the addition of hydrogen (step (4) above) is a key step in the process.

(iii) The kinetic saturation effect of increasing the olefin concentration (Figure 3) indicates that olefin attachment as in reaction (2) is a pre-equilibrium step that preceeds the rate-determining step. Hence, an increase in olefin concentration has an influence on the rate of catalysis only up to a limiting effect, and further increases in olefin concentration lead to no further increase in catalysis rate.

(iv) The temperature dependence of the hydrogenation rate constants indicates a single active catalytic species, and the negative value of $\Delta S^{\neq}$ indicates that the rate determining step involves a considerable entropic restriction, *i.e.* a concerted process.

(v) The deuterium isotope effect (1.47) indicates that H_2 or D_2 cleavage occurs at the rate determining step.

(vi) The addition of excess PPh_3 suppresses the catalysis, presumably by driving equilibrium (1) above to the left.

(d) Terminal alkenes are thought to be more readily reduced than internal ones because of the steric factors experienced by the latter, when brought close to the bulky triphenylphosphine groups. The authors argue in favor of a concerted, four-centered transition state for step (3) above.

30-3C

(a) Two routes to the dihydride were described. The first is simply protonation of the hydride with a nonmineral acid:

$$HCo[P(OCH_3)_3]_4 + H^+ \rightarrow H_2Co[P(OCH_3)_3]_4^+$$

Also they reported direct reaction with H_2:

$$Co[P(OCH_3)_3]_4^+ + H_2 \rightarrow H_2Co[P(OCH_3)_3]_4^+$$

(b) The *cis*-dihydride may be drawn in one of two ways depending on the H-H bond order in the adduct:

dihydride H_2 adduct

It is principally NMR evidence that indicates that this is a *cis* adduct, but the obviously unimolecular nature of the elimination of H_2 from the cobalt atom also is consistent with the *cis* geometry. See footnote 6 in the article.

(c) The most impressive evidence for this is the experiment in which the dihydride cation was exposed to D_2, with no observable formation of HD. Also, some exchange of D_2 and H_2 to give HD was observed in polar solvent (acetonitrile, equation (2)).

(d) Methane elimination (which occurred much faster than elimination of H_2, and which, unlike the H_2 elimination, was irreversible) was studied by protonation of an alkyl cobalt compound:

$$CH_3Co[P(OCH_3)_3]_4 + H^+ \rightarrow CH_4 + Co[P(OCH_3)_3]_4^+$$

and by methylation of the hydrido complex:

$$HCo[P(OCH_3)_3]_4 + (CH_3)_3O^+ \rightarrow CH_4 + (CH_3)_2O + Co[P(OCH_3)_3]_4^+$$

30-5C

(a) The reactant is electronically unsaturated and readily adds other ligands:

$Os(CO)_2(PPh_3)_2$

Os,	8 electrons
2 x CO,	4 electrons
2 x PPh$_3$,	4 electrons
total	16 electrons

The addition of formaldehyde to give the η^2 ligand shown bound to Os in Figure 1 gives an 18-electron substance:

$Os(CO)_2(PPH_3)_2(\eta^2-CH_2O)$

Os,	8 electrons
2 x CO,	4 electrons
2 x PPh$_3$,	4 electrons
η^2-CH$_2$O,	2 electrons
total	18 electrons

The thermal decomposition of this formaldehyde complex to give the hydrido formyl may be regarded as an oxidation of Os to give two anionic, two-electron donor ligands, H^- and HCO^-:

$Os(CO)_2(PPh_3)_2(H)(HCO)$

Os^{II},	6 electrons
2 x CO,	4 electrons
2 x PPh$_3$,	4 electrons
H$^-$,	2 electrons
HCO$^-$,	2 electrons
total	18 electrons

All subsequent reactions lead to 18-electron compounds, including reductive elimination of H_2 from the hydrido formyl to give $Os(CO)_3(PPh_3)_2$:

$Os(CO)_3(PPh_3)_2$

	Os,	8 electrons
3 x CO,		6 electrons
2 x PPh$_3$,		4 electrons
	total	18 electrons

(b) The only obvious explanation rests on the greater strength of the Os-Cl bond than the Os-H bond.

CHAPTER 31

A - Review

31-1A

The list of important metals is extensive, and it will surely grow as new discoveries are made. For now we can list the transition metals V, Cr, Mn, Fe, Co, Ni, Cu, W, and Mo, as well as the non-transition metals Na, K, Mg, Ca, and Zn.

31-3A

Light energy is absorbed by the porphine ring, and the magnesium ion serves in at least two capacities to aid in the transfer of this energy into the redox cascade. First, magnesium provides the chlorophyll with a rigid structure so that the photochemical energy is not uselessly lost in undesireable vibrational excitations. Second, magnesium enhances the interconversion from the initial excited singlet electronic state to a more long-lived excited triplet state. It thereby increases the efficiency with which this energy can be used in redox processes.

31-5A

Hemoglobin and myoglobin are heme proteins that serve as reversible oxygen binding substances in the respiration process. They each contain the heme group, but hemoglobin contains four heme units, whereas myoglobin contains only one. Hemoglobin is therefore able to exhibit the cooperativity effects in O_2 binding (Figure 31-3) that are necessary to its functioning as a carrier of oxygen. Myoglobin, on the other hand, functions without the effects of cooperativity, to store oxygen in the muscles, for release upon demand. Hemoglobin also serves to transport CO_2 back to the lungs.

31-7A

As discussed in Section 31-5, these are iron-sulfur proteins that contain clusters of two or more iron atoms, each coordinated by sulfur atoms, some of which are from cysteine residues of the protein backbone. Thus, there are two types of sulfur atoms, those of a cysteine residue and the "inorganic sulfur" atoms. The four-iron cluster contains two concentric tetrahedra, one composed of four iron atoms and a larger one composed of sulfur atoms, as shown in Figure 31-8B of the text. Each iron atom is therefore coordinated by four sulfur atoms, one of which comes from a cysteine residue. The 8Fe cluster contains two such tetrahedra loosely linked. More on this is available in the supplementary readings. Note that the three-iron clusters (Structure 31-III) are fragments of the four-iron clusters (Structure 31-IV).

31-9A

These are listed in Section 31-8. Compare Figures 31-11 and 31-13. In addition to the central cobalt atom (which may exist in different oxidation states), we have the roughly planar macrocyclic corrin ring ligand system, an axial base ligand side chain, and the sixth ligand, nominally cyanide (for cyanocobalamin, vitamin B_{12}) or 5'-deoxyadenosyl (shown in Figure 31-12). The distinction between the various oxidation states is the following: Co^{III} (vitamin B_{12}), Co^{II} (vitamin B_{12r}), and Co^{I} (vitamin B_{12s}).

31-11A

This is nitrogen fixation, the reduction of dinitrogen to ammonia, as discussed in Section 31-10.

C - Questions from the Literature of Inorganic Chemistry

31-1C

(a) The significance of this scrambling is that it supports the suggestion as shown in Figure 2 of this article, that the 5'-deoxyadenosyl radical ($ADCH_2 \cdot$) abstracts a hydrogen atom from the substrate (Figure 2, step ii), giving $ADCH_3$, in which all hydrogen atoms are equivalent. Subsequent reformation of $ADCH_2 \cdot$ in step iv of Figure 2 can be accomplished using any one of these three equivalent hydrogen atoms, and this leads to complete and uniform redistribution of any deuterium label that was originally present only in the substrate.

(b) The intermediate B_{12r} is a radical that can be detected by spectroscopic means. Reactions such as (7) and (10) are, therefore, reductions of Co^{III} (d^6) to Co^{II} (d^7).

(c) These factors are discussed at length in the paper, and include the following. Electronic factors are well illustrated by the data graphed in Figure 3, involving a system designed so that steric factors are constant. Thus, in a series of bis(dimethylglyoximato) complexes with various *trans*-L bases, the base strength of L is seen to influence the dissociation energy (D_{Co-R}) of the *trans* $C_6H_5(CH_3)HC-$ alkyl group. The data of Table 2 reveal the influence of steric factors on the value of D_{Co-R}. As shown in Figure 4, values of D_{Co-R} decrease with the growing steric effects provided by phosphine ligands with increasing cone angles. The X-ray structural data of Table 3 indicate that there is a lengthening of the Co-R bond with increased steric factors associated either with R or with L. This sort of steric crowding is also seen in coenzyme B_{12}. In fact, the close contacts in coenzyme B_{12} between the 5'-deoxyadenosyl group and the corrin ring atoms is the likely reason that the enzyme is so readily able to induce Co-C bond breaking. As Halpern has pointed out, it is this steric crowding between the alkyl ligand and the corrin ring that allows such a modest distortion to accomplish the required Co-C bond activation, giving the coenzyme its utility as a free radical precursor.

(d) Each has a roughly six-coordinate cobalt atom that undergoes facile redox changes between Co^{III}, Co^{II}, and Co^I. Each contains a substitutable sixth ligand and either a nitrogen donor atom or an alkyl ligand. Each contains a roughly planar, tetradentate, macrocyclic equatorial ligand. Furthermore, the cobaloximes and the "saloph" derivatives undergo reactions that are completely analogous to those shown for cobalamin in equations 31-8.2 through 31-8.4 of the text. Compare also Figures 31-11 and 31-13 of the text.

Given these structural similarities, (and considering the comparison of structures 1 and 2 with Figure 1 in the article), it is not surprising that these model compounds correctly mimic the redox properties of the coenzyme, and that they have cobalt-carbon bond dissociation energies that are comparable to those of the coenzyme.

(e) There is an obvious formal resemblance between the sort of $M-O_2$ bond dissociation reactions represented by equations (23) and (24) and the $M-R$ bond dissociations that accompany the activity of coenzyme B_{12}. This is substantiated by a comparison of Figures 4 and 6. Thus, whereas myoglobin and hemoglobin serve as reservoirs of reversibly-bound dioxygen, coenzyme B_{12} serves as a reservoir of free radicals.

31-3C

(a) This chemistry is of direct importance to that of the molybdenum nitrogenase enzymes because these model systems have been shown to undergo successive reductions of N_2 to NH_3. Most of the individual steps in the reduction sequence have been modeled with specific compounds and reactions. For instance, all intermediates in sequence (5), except (v), have been modeled by stable complexes. This sequence thus represents a plausible step-wise understanding of the action of nitrogenase enzymes.

(b) The reactions that gave intermediate reduction products were those of the complexes containing bis(chelate)-type phosphine ligands. Thus, the trans-$M(N_2)_2(dppe)_2$ complexes, where dppe represents a chelating diphosphine, are protonated by halogeno acids to give complexes of the ligands $-N_2H$, $-N_2H_2$, and untimately, $-NNH_2$. Complete reduction to NH_3 was not observed. It is a major point of this article that with the use of oxoacids instead of halogeno acids, or with the use of monodentate phosphines instead of chelating phosphines, one obtains high yields of NH_3 rather than intermediate reduction products such a $-N_2H$ or $-N_2H_2$.

(c) The complexes containing two bidentate dppe ligands are protonated to give only intermediate reduction products, *viz.* those mentioned in the answer to (b) above. In contrast, complexes containing four monodentate PMe_2Ph ligands in place of two dppe ligands are protonated to give full reduction to NH_3. These authors argue that it is the ready replacement of one or more PMe$_2$Ph ligands by anionic π donors such as sulfate, hydrogen sulfate, or phosphate ligands that allows further reduction to NH$_3$. This should enhance reduction beyond the -N$_2$H$_2$ stage because anionic, π donor, oxygen-containing ligands such as sulfate will release π electron density into the central metal, thereby enhancing the metal's ability to transfer negative charge to the N$_2$H$_2$ ligand. This is also consistent with the observation that oxo acids promote the reduction to ammonia better than halogeno acids. The argument requires that the chelating diphosphine ligands are not readily replaced by oxoanions under the reaction conditions.

(d) In scheme (5), the cleavage of the N-N bond takes place in the step from (v) to (vi). This is the step shown in equation (4) of the article. This is likely to occur only in cases where the metal can undergo the extra necessary increase in oxidation state.

(e) The formal increase in oxidation state that is necessary in sequence (5) is that from a metal complex that is zero valent to one of oxidation state M^{VI}. This extreme change of oxidation state is not likely necessary in the enzymic system, because electrons are no doubt passed to the molybdenum atom from iron-sulfur clusters that are part of the enzyme.

CHAPTER 32

A - REVIEW

32-1A

Liquids and gases are almost always composed of molecules (or atoms in the case of the noble gases) that are only loosely associated. (The obvious exceptions are molten salts and liquids with a high degree of hydrogen bonding, as discussed in Chapters 7 and 9.) Molecular liquids and gases are characterized by a high degree of disorder. There are many solids that are also molecular in their compositions, and these adopt structures in which ordered arrays of small molecules touch each other loosely (*i.e.* have only van der Waals contacts) in the crystal.

Apart from these, there are the vastly different nonmolecular solids. In nonmolecular materials, the atoms and ions make very strong contacts, and there are bonding interactions that extend throughout a solid sample. The nonmolecular solids include simple ionic compounds, as discussed in Chapter 4. They also include inorganic solids such as polymers, metals, alloys, and infinite covalent materials such as silicon and the numerous ceramics. These inorganic solids have interesting mechanical, electrical, magnetic and optical properties that can be understood only from the standpoint of a continuous solid array.

32-3A

(a) The shake and bake method involves heating a mixture of two or more finely divided powders in an inert, sealed container. It is simple and widely used, but gives products of unpredictable stoichiometries and sometimes of poor homogeneity. The method can be improved, in cases, by the use of a flux.

(b) Vapor phase transport (illustrated for ZnS in Figure 32-1) is used to transform amorphous solids into crystalline ones. It differs from simple sublimation in the use of a transport agent (I_2 in Figure 32-1).

(c) A flux is an aid to melting and also serves to facilitate the reaction of solid mixtures at high temperatures. It is not consumed by the reaction in a stoichiometric sense, but provides shorter reaction times, often at lower temperatures than are necessary in the absence of the flux. The use of a flux constitutes an important embellishment to the shake and bake method.

(d) A hydrothermal synthesis is one that employs water, often in the supercritical state. In such cases, it is a synthesis that is conducted in a sealed container above the triple point temperature of water. In other cases, water is used, but at a temperature below the triple point. The synthesis of hydrated mordenite is an example. The method is particularly useful for the preparation of hydrated substances. Otherwise, the water must be driven off in a subsequent step if anhydrous materials are desired.

32-5A

This is given in Figure 32-1.

32-7A

As shown in Figure 32-4, the valence shells in the alkali metals give rise to two bands. For metallic sodium, the bands overlap, as in part (b) of the figure.

32-9A

A band gap is that energy between the highest energy of one band and the lowest energy of the next, that is, an energy region where adjacent bands are completely nonoverlapping. This is illustrated in Figure 32-6, where the density of states is plotted versus energy, the band gap being the energy between the uppermost edge of the s band and the lowermost edge of the p band.

32-11A

An insulator is a nonconductor. Nonconductors are substances in which there are only either completely filled or completely empty bands, separated from one another by a large band gap. A semiconductor has a small conductance that increases with increasing temperature. Semiconductors are substances in which the band gap is small compared to insulators. This is shown in Figure 32-8(a) for the special case of an intrinsic semiconductor.

32-13A

An **n**-type semiconductor is an extrinsic (that is, doped) semiconductor in which a small amout of material has been added to a host such that a filled band is placed close in energy to the upper, unfilled band of the host. This is illustrated in Figure 32-8(b). An example is the doping of silicon with a small amount of arsenic.

32-15A

It is only at T = 0 K that defects in solids are not required by thermodynamics. This is illustrated in Figure 32-9.

32-17A

Vacancies are point defects that are sometimes called Shottky defects. They are lattice points in a solid where an atom is missing. In ionic substances, the number of cation vacancies is matched by the number of anion vacancies, so as to preserve electroneutrality. The other type of point defect is the interstitial defect.

32-19A

(a) TiO

(b) Wurtzite

(c) CdS, strongly heated

(d) These are the higher oxides of titanium, vanadium, molybdenum, and tungsten, e.*g.* WO_3.

32-21A

Difussion takes place when defects exist, because, when holes are present, adjacent atoms can move into them.

32-23A

NASICON is an acronym for "sodium superionic conductor." It is the substance whose formula is $Na_3Zr_2PSi_2O_{12}$. The sodium ions reside in a small fraction of the sites in tunnels formed by corner-sharing ZrO_6 octahedra and PO_4 or SiO_4 tetrahedra. Under the influence of an electric field, the sodium ions can migrate from one site to another, giving rise to conductivity in the solid.

32-25A

Nonstoichiometric hydrides have formulas that fall short of the limiting composition MH, MH_2, or MH_3. Examples are: $NbH_{0.7}$, $ZrH_{1.6}$, and $LuH_{2.2}$.

32-27A

When alkali metals are intercalated (inserted) into the layers of the graphite structure, they transfer their valence electrons to the graphite layers and thereby become cations. The resulting compound is obviously not stoichiometric.

32-29A

Kaolins are silicate clays. They are used in the manufacture of many ceramic materials. For instance, porcelain is made from kaolin mixed with a finely powdered feldspar, which serves as a binder when the material is cooled.

32-31A

A glass is a ceramic material in a vitreous state. That is, a glass is a solid that has been obtained by cooling a liquid so that crystallization into an ordered state cannot occur. A glass is thus a disordered (vitreous) solid. A glass temperature (sometimes also called a glass transition temperature) is the temperature at which a super-cooled liquid will solidify to form a glass rather than a crystalline solid. This is illustrated in Figure 32-15.

32-33A

Superconductors include mercury (the one discovered first) and other metals and alloys such as Nb_3Ge, plus modern materials such as $PbMo_6S_8$ and other Chevrel phases, $Ba_{0.8}Y_{1.2}CuO_x$, $Tl_2Ba_2Ca_2Cu_3O_{10}$, and $YBa_2Cu_3O_7$.

32-35A

Some of the iron atoms in "FeO" are in the Fe^{III} oxidation state, and this requires, overall, fewer iron atoms to balance the charge of the O^{2-} ions.

32-37A

This is illustrated in Figure 32-16. A low external voltage applied in a direction perpendicular to the junction will cause current to flow in one direction (from the negative terminal towards the positive terminal) but not in the other direction.

32-39A

Superconductivity is suppressed by the application of a strong magnetic field. In many materials, the effect has an abrupt onset at some minimum field strength, H_c, whose value is characteristic of the material.

B - Additional Exercises

32-1B

Extrinsic semiconductors have the advantage that the amount and the nature of the added material can be varied, thereby adjusting the properties of the semiconductor. Consider, for example, silicon, which can be doped with varying amounts of arsenic. By altering the nature of the added substance, the size of the gaps in Figure 32-8(b) can be altered.

32-3B

This arises because of the creation of color center defects in the sample. The defects are created by loss of sulfur atoms from the sample at high temperatures.

32-5B

The vacant sites that one expects to find are closed up by the development of a shear plane, as diagrammed in Figure 32-10 and shown in Figure 32-11.

32-7B

This is not altogether true, as discussed in Section 32-7. The compound is one of the high-temperture superconductors, and Cooper pairs can persist only at low temperatures.

32-9B

The existence of a defect in a crystal lattice disrupts the normal stability associated with the lattice. Compounds with large lattice energies should, therefore, be less prone to defects. This is confirmed by the general observation that compounds that have low densities are especially prone to defects.